PRESI
LOCOMOTIVES
OF BRITISH RAILWAYS

FOURTEENTH EDITION

Robert Pritchard & Peter Hall

Published by Platform 5 Publishing Ltd., 3 Wyvern House, Sark Road, Sheffield. S2 4HG, England.

Printed in England by Charlesworth Press, Flanshaw Way, Flanshaw Lane, Wakefield, West Yorkshire, WF2 9LP.

ISBN 978 1 902336 84 8

CONTENTS

Foreword to the Fourteenth edition **4**
Acknowledgements and contact details **4**
Introduction **5**
General Notes **5**
1. Steam Locomotives **6**
1.1. Great Western Railway & Absorbed Company Locomotives 7
1.2. Southern Railway & Constituent Company Locomotives 22
1.3. London Midland & Scottish Railway & Constituent Company Locomotives 34
1.4. London & North Eastern Railway & Constituent Company Locomotives 50
1.5. British Railways Standard Steam Locomotives 64
1.6. War Department Locomotives 69
1.7. United States Army Transportation Corps Steam Locomotives 73
1.8. New build steam locomotives 74
1.9. Replica Steam Locomotives 77
2. Diesel Locomotives **80**
2.1. London Midland & Scottish Railway Diesel Locomotives 81
2.2. Southern Railway Diesel Locomotives 82
2.3. British Railways Diesel Locomotives 83
2.4. Experimental Diesel Locomotives 111
2.5. Civil Engineers Diesel Locomotives 112
3. Electric Locomotives **113**
3.1. Pre-Nationalisation Design Electric Locomotives 113
3.2. British Railways Electric Locomotives 114
4. Gas Turbine Vehicles **120**
5. Multiple Unit Vehicles **121**
5.1. Steam Railmotor 122
5.2. Petrol-Electric Autocar 122
5.3. GWR Diesel Railcars 122
5.4. British Railways DMUs 124
5.5. Four-wheeled Diesel Railbuses 135
5.6. Diesel Electric Multiple Units 139
6. Electric Multiple Unit Vehicles **141**
6.1. Southern Railway EMU Stock 141
6.2. Pullman Car Company EMU Stock 143
6.3. LMS & Constituents EMU Stock 144
6.4. LNER & Constituents EMU Stock 145
6.5. British Railways EMU Stock 146
Appendix I. List of Locations **155**
Preservation Sites & Operating Railways 155
Appendix II. Abbreviations Used **158**
Appendix III. Private Manufacturer Codes **159**

Front Cover Photograph: BR 2MT 2-6-0 78019 leaves Bolton Abbey for Embsay on 24 September 2009. **Brian Dobbs**

Left: A pair of LMS 5MT 4-6-0s, 45231 and 45407, climb towards Blaenau Ffestiniog at Blaenau Dolwyddelan on 2 May 2009 with the "The Welsh Mountaineer" charter. **Terry Eyres**

Back Cover Photograph (top): D1062 "WESTERN COURIER" approaches Williton on the West Somerset Railway with the 16.00 Bishops Lydeard–Minehead on 12 June 2010. **Lindsay Atkinson**

Back Cover Photograph (bottom): The Keighley & Worth Valley Railway Class 108 DMU, 51565+50928, leaves Keighley with the 10.00 to Oxenhope on 13 April 2010. **Andrew Wills**

FOREWORD TO THE FOURTEENTH EDITION

Welcome to the fourteenth edition of Platform 5's guide to Preserved Locomotives of British Railways.

For this edition we have introduced a new section covering New Build Steam Locomotives. There are many projects for "new" locomotives now under way, all at differing stages of progress. Many have a long way to go, but one only has to look at the amazing 17-year project to build A1 60163 "TORNADO", now a regular performer on the main line, to see what can be achieved. All these locomotives can all now be found within the sections for New Build Steam Locomotives or Replica Locomotives.

Another significant achievement has been the construction of Lynton & Barnstaple 2-6-2T "LYD" at the Ffestiniog Railway's Boston Lodge works. Meanwhile, at the Llangollen Railway, the restoration of Steam Railmotor No. 93 has been completed, and this fascinating machine is now based at Didcot Railway Centre.

Another new addition to this edition is "Dunrobin", an 0-4-4T built for the Duke of Sutherland in 1895 and used on the Highland Railway. This loco was exported to Canada in 1965 but repatriated to Britain just before this edition closed for press.

Foreign preserved steam locomotives resident in Great Britain are no longer included in this book.

The growth in diesel preservation has continued, with further diesel locomotives being secured for preservation. It should be noted that for some diesel and electric locomotives that are currently considered to be "preserved", this status is not irreversible. Several locomotives that have appeared in previous editions of this book are no longer included here, some having returned to commercial operation and others having been used as a source of spares before being scrapped.

It is our aim that all former BR diesel and electric locomotives will be listed either in this book if classed as "Preserved" or in the annual Platform 5 "Locomotives" pocket book or "Locomotives & Coaching Stock" bound volume. A small number of locomotives may be found in both books if they are truly preserved locomotives but are also main line registered, for example Deltic D9000 "ROYAL SCOTS GREY" or 86259.

Peter Fox 1942–2011

The first 13 editions of this book were either edited or co-edited by Peter Fox, Managing Director of Platform 5 Publishing, with the first edition appearing in 1983. Peter, who was also editor of our magazine Today's Railways UK, sadly lost his battle with heart disease in February 2011, but we are pleased to keep his hard work going in the form of this book as well as the other books and magazines published by Platform 5. In May 2011 Grand Central Railway Company honoured Peter by naming one of its HST power cars "PETER FOX 1942–2011 – PLATFORM 5" on Platform 5 at Sheffield station.

ACKNOWLEDGEMENTS & CONTACT DETAILS

Thanks are given to all those individuals and railways who have knowingly and unknowingly assisted in the updating of this book. In particular the authors would like to thank the many readers who have contacted us with amendments and comments relating to the thirteenth edition of this book. We would also like to thank Paul Abell and Ian McLoughlin for their help in the production of this publication.

As the authors are not infallible, they would welcome notification of any corrections or updates to this book of which readers have first-hand knowledge.

Please send comments or amendments to Robert Pritchard at the Platform 5 address on the title page or by e-mail to robert@platform5.com (telephone 0114 255 2625).

Robert Pritchard & Peter Hall. June 2011.

UPDATES

Updates to this book are published in Platform 5's monthly magazine, **Today's Railways UK**, which is the only magazine to carry official Platform 5 stock changes. **Today's Railways UK** is available from good newsagents or on direct subscription (see inside covers of this book).

INTRODUCTION

This book contains details of preserved locomotives and multiple unit stock which has been in the ownership of, built to the designs of, or operated under the jurisdiction of, the British Railways Board, its constituents or its descendents, in revenue earning service or on the "main line". Locomotives solely used in workshops and depots are excluded as are those solely built for export. Only vehicles no longer in the ownership of the descendents of the British Railways Board or the commercial railway companies are included.

Also included are War Department "Austerity" design steam locomotives and those steam locomotives built for the USATC which are currently resident in Great Britain.

This book does not include locomotives or multiple unit vehicles from overseas railways preserved in Great Britain.

The book is updated to information received by early June 2011.

GENERAL NOTES

This book has been divided into several main categories namely steam locomotives, diesel locomotives, electric locomotives, gas turbine vehicles and multiple unit stock. Further details applicable to each category being given in the introductory paragraph of each section.

Notes regarding technical details of the various locomotives and multiple units can be found in the introductory paragraphs for each section. A few details are, however, consistent for each section, these being:

NUMBERS

All numbers carried at various times by locomotives are shown, except in the case of temporary identity changes for filming and similar events and numbers carried by locomotives when in industrial use unless they are still carried. For each railway, numbers are given in chronological order. Details are also given of the current identity if it had not previously been carried.

NAMES AND CLASS NAMES

Names bestowed after preservation are shown in inverted commas and are only shown if still carried. Official names and class names are shown without inverted commas and these are shown irrespective of whether they are still carried.

If more than one official name has been carried then the most recent is shown with those previously carried given as a footnote. If however an earlier official name is carried then this is shown and details of other official names carried are given as a footnote.

LOCATIONS

The location where the locomotive or multiple unit is normally to be found is given. Fuller details of locations in Great Britain including OS grid references are provided under "List of Locations". It is not uncommon for locomotives and multiple units to visit other locations for operation, display or mechanical attention. Where such visits are of a long-term nature, the locomotive or multiple unit is shown as being at its host site.

(N) denotes a locomotive or multiple unit vehicle that forms part of the National Collection.

GAUGE

All locomotives and multiple units are standard gauge unless stated otherwise.

BUILD DETAILS

For each locomotive the builder, works number (if any) and year of build are given. Private builder codes will be found in Appendix III.

1. STEAM LOCOMOTIVES

Almost from the birth of the railways to the 1960s, steam was the principal form of propulsion, its rapid decline in the 1960s being the impulse for much of the preservation movement. Steam locomotives are arranged generally in numerical order of the British Railways number, except that very old locomotives which did not receive numbers in the series pertaining at nationalisation in 1948 are listed at the end of each pre-nationalisation company section.

A select number of steam locomotives are permitted to work steam specials on the national railway network. Such locomotives invariably spend long periods away from their home bases undertaking such duties. Other locomotives may also spend periods away from their home bases as guests of other locations.

WHEEL ARRANGEMENT

The Whyte notation is used for steam locomotives and diesel shunters with coupled driving wheels. The number of leading wheels are given, followed by the number of driving wheels and then the trailing wheels. Suffixes are used to denote tank locomotives as follows: T – side tank, PT – pannier tank, ST – saddle tank, WT – well tank. For example 2-6-2T.

DIMENSIONS

These are given in imperial units for steam locomotives as follows:

Boiler pressure.	In pounds force per square inch (lbf/sq. in.).
Cylinders	Cylinder dimensions. The diameter is given first followed by the stroke. (I) indicates two inside cylinders, (O) two outside cylinders, (V) two vertical cylinders, (3) three cylinders (two outside and one inside) and (4) four cylinders (two outside and two inside).
Wheel diameters	These are given from front to back, i.e. leading, driving, trailing.
Weights	These are given in full working order.

TRACTIVE EFFORT

This is given at 85% boiler pressure to the nearest 10 lbf for steam locomotives. Phillipsons formula has been used to calculate these as follows:

$$TE = \frac{0.85d^2snp}{2w}$$

where TE = tractive effort in pounds force;
d = cylinder diameter (inches);
s = piston stroke (inches);
n = number of cylinders;
p = maximum boiler pressure (lb/sq. in.);
w = new driving wheel diameter (inches).

BRAKES

Steam locomotives are assumed to have train vacuum brakes unless otherwise stated.

VALVE GEAR

Unless stated otherwise valve gear on steam locomotives is assumed to be inside the locomotive, except Walschaerts and Caprotti gears which are assumed to be outside. Exceptions are LMS-design 5MT 4-6-0 44767 with outside Stephenson valve gear, LNER (ex-GER) N7 0-6-2T No. 69621 with inside cylinders and Walschaerts valve gear, and various 4-cylinder GWR 4-6-0s with inside Walschaerts valve gear and rocking shafts for the outside cylinders.

1.1. GREAT WESTERN RAILWAY AND ABSORBED COMPANIES' STEAM LOCOMOTIVES

GENERAL

The GWR was the only one of the big four companies which was essentially an existing company. A number of smaller companies were absorbed at the grouping. These were virtually all in Wales and included the Cambrian Railways, Cardiff Railway, Rhymney Railway and Taff Vale Railway. Prior to 1923 the GWR had also absorbed smaller concerns at various dates.

NUMBERING & CLASSIFICATION SYSTEM

The locomotives of the absorbed companies were given the lower numbers and GWR classes the higher numbers. Instead of arranging classes in blocks, the GWR adopted a system whereby the second digit remained constant within a class, eg the 0-6-2Ts numbered 5600–99 continued with 6600–99. Sometimes earlier numbers were filled in, e.g. 5101–99 continued with 4100–99. Classes were always denoted by the number of the first member of the class to be built, which was not always the lowest number in the series. GWR locos were not renumbered by BR on nationalisation.

POWER CLASSIFICATION & ROUTE RESTRICTION SYSTEM

The GWR adopted a power classification letter code system which ranged from A to E in ascending order of power. Certain small locomotives which were below group A were said to be unclassified and the "Kings" were classed as "special" being higher than "E". BR power classifications are also shown in brackets in this section.

The power classification letter was shown on the cabside on a coloured spot showing the route restriction. In ascending order of restriction these were as follows: Yellow, Blue, Red, Double Red. Where no restriction is specified, locomotives were unrestricted.

▶ After a very varied career, during which it was actually taken over by the Great Western Railway on two different occasions 25 years apart, North Pembrokeshire & Fishguard Railway 0-6-0ST 1378 "MARGARET" is on display at the Scolton Manor Museum, near Haverfordwest, where it is seen on 23 August 2010. **Ian McLoughlin**

CORRIS RAILWAY 0-4-2ST

Built: 1878 by Hughes Locomotive & Tramway Engine Works Ltd. as 0-4-0ST. Rebuilt 1900 to 0-4-2ST.
Gauge: 2′ 3″.
Boiler Pressure: 160 lbf/sq in.
Wheel Diameters: 2′ 6″, 10″.
Valve Gear: Stephenson.
Weight: 9 tons.
Cylinders: 7″ x 12″ (O).
Tractive Effort: 2670 lbf.

3	"SIR HAYDN"	Talyllyn Railway	HLT 323/1878 reb. 1901

CORRIS RAILWAY 0-4-2ST

Built: 1921.
Boiler Pressure: 160 lbf/sq in.
Wheel Diameters: 2′ 0″, 1′ 4½″.
Valve Gear: Hackworth.
Gauge: 2′ 3″.
Weight: 8 tons.
Cylinders: 7″ x 12″ (O).
Tractive Effort: 3330 lbf.

4	"EDWARD THOMAS"	Talyllyn Railway	KS 4047/1921

VALE OF RHEIDOL 2-6-2T

Built: 1923. GWR development of an older design built 1902.
Gauge: 1′ 11½″.
Boiler Pressure: 165 lbf/sq in.
Wheel Diameters: 2′ 0″, 2′ 6″, 2′ 0″.
Valve Gear: Walschaerts. Piston valves.
Weight: 25 tons.
Cylinders: 11½″ x 17″ (O).
Tractive Effort: 10510 lbf.

7	OWAIN GLYNDŴR	Vale of Rheidol Railway	Swindon 1923
8	LLYWELYN	Vale of Rheidol Railway	Swindon 1923
9	PRINCE OF WALES	Vale of Rheidol Railway	Swindon 1923

No. 12 4wT

Built: 1926. Sentinel vertical-boilered geared locomotive. Returned to manufacturer after 3 months service.
Boiler Pressure: 275 lbf/sq in.
Wheel Diameter: 2′ 6″.
Valve Gear: Rotary cam.
Weight: 20 tons.
Cylinders: 6″ x 9″ (I).
Tractive Effort: 7200 lbf.

GWR	*Present*			
12	49	"No. 2 ISEBROOK"	Buckinghamshire Railway Centre	S 6515/1926

TAFF VALE RAILWAY CLASS 02 0-6-2T

Built: 1899 by Neilson, Reid. Sold by GWR 1926. 9 built.
Boiler Pressure: 160 lbf/sq in.
Wheel Diameters: 4′ 6½″, 3′ 1″.
Valve Gear: Stephenson. Slide valves.
Power Classification: B.
Weight: 61.5 tons.
Cylinders: 17½″ x 26″ (I).
Tractive Effort: 19870 lbf.
Restriction: Blue.

GWR	*TVR*		
426	85	Keighley & Worth Valley Railway	NR 5408/1899

TAFF VALE RAILWAY CLASS 01 0-6-2T

Built: 1894–97. Survivor sold by GWR 1927. 14 built.
Boiler Pressure: 150 lbf/sq in.
Wheel Diameters: 4′ 6½″, 3′ 8¾″.
Valve Gear: Stephenson. Slide valves.
Power Classification: A.
Weight: 56.4 tons.
Cylinders: 17½″ x 26″ (I).
Tractive Effort: 18630 lbf.
Restriction: Yellow.

GWR	*TVR*		
450	28	Llangollen Railway (N)	Cardiff West Yard 306/1897

▲ Built for the Corris Railway in 1878, 0-4-2ST No. 3 "SIR HAYDN" is still a regular performer on the Talyllyn Railway. On 9 May 2011 it arrives at Dolgoch with the 15.00 Tywyn Wharf–Nant Gwernol.
Robert Pritchard

▼ 1400 Class 0-4-2T 1450 is seen at Didcot Railway Centre on 3 May 2010. **Alisdair Anderson**

▲ 2251 Class 0-6-0 3205 runs alongside the River Dart at Hood Bridge on the South Devon Railway with a service for Totnes Riverside on 25 April 2011. **Stacey Thew**

▼ Newly restored in BR Black, 2884 Class 2-8-0 3850 heads the Great Central mineral wagons at Swithland Sidings on 7 February 2011. **Hugh Ballantyne**

PORT TALBOT RAILWAY 0-6-0ST

Built: 1900/01. Survivor sold by GWR 1934. 6 built.
Boiler Pressure: 160 lbf/sq in.
Wheel Diameter: 4′ 0½″.
Valve Gear: Stephenson. Slide valves.
Power Classification: A.
Weight: 44 tons.
Cylinders: 16″ x 24″ (I).
Tractive Effort: 17 230 lbf.
Restriction: Yellow.

GWR	*PTR*		
813	26	Severn Valley Railway	HC 555/1901

WELSHPOOL & LLANFAIR RAILWAY 0-6-0T

Built: 1903. 2 built.
Boiler Pressure: 150 lbf/sq in.
Wheel Diameter: 2′ 9″.
Valve Gear: Walschaerts. Slide valves.
Gauge: 2′ 6″.
Weight: 19.9 tons.
Cylinders: 11½″ x 16″ (O).
Tractive Effort: 8180 lbf.

BR	*GWR*	*W&L*			
822	822	1	THE EARL	Welshpool & Llanfair Light Railway	BP 3496/1903
823	823	2	COUNTESS	Welshpool & Llanfair Light Railway	BP 3497/1903

823 originally named THE COUNTESS. Name altered to COUNTESS by GWR.

POWLESLAND & MASON 0-4-0ST

Built: 1903–06. Survivor sold by GWR 1928 for industrial use.
Boiler Pressure: 140 lbf/sq in.
Wheel Diameter: 3′ 6″.
Valve Gear: Stephenson. Slide valves.
Weight: 24.85 tons.
Cylinders: 14″ x 20″ (O).
Tractive Effort: 11 110 lbf.

GWR	*P&M*		
921	6	Snibston Discovery Park	BE 314/1906

CARDIFF RAILWAY 0-4-0ST

Built: 1898. Rebuilt Tyndall Street Works 1916.
Boiler Pressure: 160 lbf/sq in.
Wheel Diameter: 3′ 2½″.
Valve Gear: Hawthorn-Kitson.
Weight: 25.5 tons.
Cylinders: 14″ x 21″ (O).
Tractive Effort: 14540 lbf.

GWR	*CARR*		
1338	5	Didcot Railway Centre	K 3799/1898

ALEXANDRA DOCKS & RAILWAY COMPANY 0-4-0ST

Built: 1897. Rebuilt Swindon 1903. Sold by GWR 1932 for industrial use.
Boiler Pressure: 120 lbf/sq in.
Wheel Diameter: 3′ 0″.
Valve Gear: Stephenson. Slide valves.
Weight: 22.5 tons.
Cylinders: 14″ x 20″ (O).
Tractive Effort: 11 110 lbf.

GWR	*AD*		
1340 TROJAN	TROJAN	Didcot Railway Centre	AE 1386/1897

1361 CLASS 0-6-0ST

Built: 1910. Churchward design for dock shunting. 5 built (1361–5).
Boiler Pressure: 150 lbf/sq in.
Wheel Diameter: 3′ 8″.
Valve Gear: Allan.
Power Classification: Unclassified (0F).
Weight: 35.2 tons.
Cylinders: 16″ x 20″ (O).
Tractive Effort: 14840 lbf.

1363	Didcot Railway Centre	Swindon 2377/1910

1366 CLASS 0-6-0PT

Built: 1934. Collett design for dock shunting. Used to work Weymouth Quay boat trains. 6 built. (1366–71).
Boiler Pressure: 165 lbf/sq in.
Wheel Diameter: 3′ 8″.
Valve Gear: Stephenson. Slide valves.
Power Classification: Unclassified (1F).
Weight: 35.75 tons.
Cylinders: 16″ x 20″ (O).
Tractive Effort: 16320 lbf.

1369	South Devon Railway	Swindon 1934

NORTH PEMBROKESHIRE & FISHGUARD RAILWAY 0-6-0ST

Built: 1878. Absorbed by GWR 1898. Sold to Gwendraeth Valley Railway in 1910. Absorbed by GWR again in 1923 but sold in March of that year to Kidwelly Tinplate Company.
Boiler Pressure: 140 lbf/sq in.
Wheel Diameter: 4′ 0″.
Valve Gear: Stephenson. Slide valves.
Weight: 30.95 tons.
Cylinders: 16″ x 22″ (I).
Tractive Effort: 13960 lbf.

GWR	*GVR*			
1378	2	MARGARET	Scolton Manor Museum	FW 410/1878

1400 CLASS 0-4-2T

Built: 1932–36. Collett design. Push & Pull fitted. Locos renumbered in 1946. 75 built.(1400–74).
Boiler Pressure: 165 lbf/sq in.
Wheel Diameters: 5′ 2″, 3′ 8″.
Valve Gear: Stephenson. Slide valves.
Power Classification: Unclassified (1P).
Weight: 41.3 tons.
Cylinders: 16″ x 24″ (I).
Tractive Effort: 13900 lbf.

1932 No.	*1946 No.*		
4820	1420	South Devon Railway	Swindon 1933
4842	1442	Tiverton Museum	Swindon 1935
4850	1450	Dean Forest Railway	Swindon 1935
4866	1466	Didcot Railway Centre	Swindon 1936

1500 CLASS 0-6-0PT

Built: 1949. Hawksworth design. 10 built (1500–09).
Boiler Pressure: 200 lbf/sq in.
Wheel Diameter: 4′ 7½″.
Valve Gear: Walschaerts. Piston valves.
Power Classification: C (4F).
Weight: 58.2 tons.
Cylinders: 17½″ x 24″ (O).
Tractive Effort: 22510 lbf.
Restriction: Red.

1501	Severn Valley Railway	Swindon 1949

1600 CLASS 0-6-0PT

Built: 1949–55. Hawksworth design. 70 built (1600–69).
Boiler Pressure: 165 lbf/sq in.
Wheel Diameter: 4′ 1½″.
Valve Gear: Stephenson. Slide valves.
Power Classification: A (2F).
Weight: 41.6 tons.
Cylinders: 16½″ x 24″ (I).
Tractive Effort: 18510 lbf.
Restriction: Uncoloured.

1638	Kent & East Sussex Railway	Swindon 1951

2251 CLASS 0-6-0

Built: 1930–48. Collett design. 120 built (2251–99, 2200–50, 3200–19).
Boiler Pressure: 200 lbf/sq in superheated.
Wheel Diameter: 5′ 2″.
Cylinders: 17½″ x 24″ (I).
Tractive Effort: 20 150 lbf.
Power Classification: B (3MT).
Weight–Loco: 43.4 tons.
–Tender: 36.75 tons.
Valve Gear: Stephenson. Slide valves.
Restriction: Yellow.

3205	South Devon Railway	Swindon 1946

▲ The ever-popular 3200 Class 4-4-0 "Dukedog" 9017 is seen near Caddaford heading for Totnes Riverside on 7 April 2011. The loco was visiting from the Bluebell Railway. **Ralph Ward**

▼ Painted in the maroon colours of London Transport, 5700 Class 0-6-0PT L99 (7715) hauls a demonstration freight at Tunbridge Wells West on the Spa Valley Railway on 24 October 2010. **Phil Barnes**

2301 CLASS DEAN GOODS 0-6-0

Built: 1883–99. Dean design. 280 built (2301–2580).
Boiler Pressure: 180 lbf/sq in superheated.
Wheel Diameter: 5′ 2″.
Cylinders: 17½″ x 24″ (I).
Tractive Effort: 18 140 lbf.
Power Classification: A (2MT).
Weight–Loco: 37 tons.
–Tender: 36.75 tons.
Valve Gear: Stephenson. Slide valves.
Restriction: Uncoloured.

2516	Steam – Museum of the Great Western Railway (N)	Swindon 1557/1897

2800 CLASS 2-8-0

Built: 1903–19. Churchward design for heavy freight. 84 built (2800–83).
Boiler Pressure: 225 lbf/sq in superheated.
Wheel Diameters: 3′ 2″, 4′ 7½″.
Cylinders: 18½″ x 30″ (O).
Tractive Effort: 35 380 lbf.
Power Classification: E (8F).
Weight–Loco: 75.5 tons.
–Tender: 43.15 tons.
Valve Gear: Stephenson. Piston valves.
Restriction: Blue.

2807	Gloucestershire Warwickshire Railway	Swindon 2102/1905
2818	National Railway Museum, York (N)	Swindon 2122/1905
2857	Severn Valley Railway	Swindon 2763/1918
2859	Llangollen Railway	Swindon 2765/1918
2861	Barry Rail Centre	Swindon 2767/1918
2873	South Devon Railway	Swindon 2779/1918
2874	Honeybourne Airfield Industrial Estate	Swindon 2780/1918

2884 CLASS 2-8-0

Built: 1938–42. Collett development of 2800 Class with side window cabs. 81 built (2884–99, 3800–64).
Boiler Pressure: 225 lbf/sq in superheated.
Wheel Diameters: 3′ 2″, 4′ 7½″.
Cylinders: 18½″ x 30″ (O).
Tractive Effort: 35 380 lbf.
Power Classification: E (8F).
Weight–Loco: 76.25 tons.
–Tender: 43.15 tons.
Valve Gear: Stephenson. Piston valves.
Restriction: Blue.

2885	Birmingham Moor Street Station	Swindon 1938
3802	Llangollen Railway	Swindon 1938
3803	South Devon Railway	Swindon 1939
3814	North Yorkshire Moors Railway	Swindon 1940
3822	Didcot Railway Centre	Swindon 1940
3845	Honeybourne Airfield Industrial Estate	Swindon 1942
3850	West Somerset Railway	Swindon 1942
3855	East Lancashire Railway	Swindon 1942
3862	Northampton & Lamport Railway	Swindon 1942

3200 CLASS "DUKEDOG" 4-4-0

Built: Rebuilt 1936–39 by Collett using the frames of "Bulldogs" and the boilers of "Dukes". 30 built (9000–29).
Boiler Pressure: 180 lbf/sq in
Wheel Diameters: 3′ 8″, 5′ 8″.
Cylinders: 18″ x 26″ (I).
Tractive Effort: 18 950 lbf.
Power Classification: B.
Weight–Loco: 49 tons.
–Tender: 40 tons.
Valve Gear: Stephenson. Slide valves.
Restriction: Yellow.

3217–9017	Bluebell Railway	Swindon 1938

3700 CLASS CITY 4-4-0

Built: 1903. Churchward design. Reputed to be the first loco to attain 100 mph hauling a Plymouth–Paddington "Ocean Mails" special in 1904.
Boiler Pressure: 200 lbf/sq in superheated. **Weight–Loco:** 55.3 tons.
Wheel Diameters: 3′ 2″, 6′ 8½″. **–Tender:** 36.75 tons.
Cylinders: 18″ x 26″ (I). **Valve Gear:** Stephenson. Piston valves.
Tractive Effort: 17 800 lbf.

BR	*GWR*			
3440	3717	CITY OF TRURO	Gloucestershire Warwickshire Railway (N)	Swindon 2000/1903

4000 CLASS STAR 4-6-0

Built: 1906–23. Churchward design for express passenger trains. 73 built (4000–72).
Boiler Pressure: 225 lbf/sq in superheated. **Weight–Loco:** 75.6 tons.
Wheel Diameters: 3′ 2″, 6′ 8½″. **–Tender:** 40 tons.
Cylinders: 15″ x 26″ (4). **Tractive Effort:** 27 800 lbf.
Valve Gear: Inside Walschaerts. Rocking levers for outside valves. Piston valves.
Power Classification: D (5P). **Restriction:** Red.

4003	LODE STAR	Steam – Museum of the Great Western Railway (N)	Swindon 2231/1907

4073 CLASS CASTLE 4-6-0

Built: 1923–50. Collett development of Star. 166 built (4073–99, 5000–5099, 7000–37). In addition one Pacific (111) and five Stars (4000/9/16/32/7) were rebuilt as Castles.
Boiler Pressure: 225 lbf/sq in superheated. **Weight–Loco:** 79.85 tons.
Wheel Diameters: 3′ 2″, 6′ 8½″. **–Tender:** 46.7 tons.
Cylinders: 16″ x 26″ (4). **Tractive Effort:** 31 630 lbf.
Valve Gear: Inside Walschaerts. Rocking levers for outside valves. Piston valves.
Power Classification: D (7P). **Restriction:** Red.

d–Rebuilt with double chimney. x–Dual (air/vacuum) brakes.

4073	CAERPHILLY CASTLE	Steam – Museum of the Great Western Railway (N)	Swindon 1923
4079	PENDENNIS CASTLE	Didcot Railway Centre	Swindon 1924
5029x	NUNNEY CASTLE	Tyseley Locomotive Works	Swindon 1934
5043d	EARL OF MOUNT EDGCUMBE	Tyseley Locomotive Works	Swindon 1936
5051	EARL BATHURST	Didcot Railway Centre	Swindon 1936
5080	DEFIANT	Buckinghamshire Railway Centre	Swindon 1939
7027	THORNBURY CASTLE	Crewe Heritage Centre	Swindon 1949
7029d	CLUN CASTLE	Tyseley Locomotive Works	Swindon 1950

5043 was named BARBURY CASTLE to 09/37.
5051 was named DRYSLLWYN CASTLE to 08/37.
5080 was named OGMORE CASTLE to 01/41.

4200 CLASS 2-8-0T

Built: 1910–23. Churchward design. 105 built (4201–99, 4200, 5200–4).
Boiler Pressure: 200 lbf/sq in superheated. **Weight:** 81.6 tons.
Wheel Diameters: 3′ 2″, 4′ 7½″. **Cylinders:** 18½″ x 30″ (O).
Valve Gear: Stephenson. Piston valves. **Tractive Effort:** 31 450 lbf.
Power Classification: E (7F). **Restriction:** Red.

4247		Flour Mill Workshop, Bream	Swindon 2637/1916
4248		Steam – Museum of the Great Western Railway	Swindon 2638/1916
4253		Pontypool & Blaenavon Railway	Swindon 2643/1917
4270		Rye Farm, Wishaw, Sutton Coldfield	Swindon 2850/1919
4277	"HERCULES"	Dartmouth Steam Railway	Swindon 2857/1920

▲ Castle Class 4-6-0 4073 "CAERPHILLY CASTLE" is seen on display at the Museum of the Great Western Railway at Swindon on 25 September 2010. **Robert Pritchard**

▼ 4500 Class 2-6-2T 4566 is based at the Severn Valley Railway. On 20 February 2010 it was pictured leaving Highley for Kidderminster. **Paul Abell**

4300 CLASS 2-6-0

Built: 1911–32. Churchward design. 342 built (4300–99 (renumbered from 8300–99 between 1944 and 1948), 6300–99, 7300–21, 7322–41 (renumbered from 9300–19 between 1956 and 1959).
Boiler Pressure: 200 lbf/sq in superheated. **Weight–Loco**: 62 tons.
Wheel Diameters.: 3′ 2″, 5′ 8″. **–Tender**: 40 tons.
Cylinders: 18½″ x 30 (O). **Valve Gear**: Stephenson. Piston valves.
Tractive Effort: 25 670 lbf.
Power Classification: D (4MT). **Restriction**: Blue.

8322–5322	Didcot Railway Centre	Swindon 1917
9303*–7325	Severn Valley Railway	Swindon 1932

In addition 5193 has been rebuilt by the West Somerset Railway as 4300 Class 2-6-0 tender locomotive No. 9351. See New Build section.

4500 CLASS 2-6-2T

Built: 1906–24. Churchward design. (§ Built 1927–29. Collett development with larger tanks). 175 built (4500–99, 5500–74).
Boiler Pressure: 200 lbf/sq in superheated. **Weight**: 57.9 tons (§ 61 tons).
Wheel Diameters: 3′ 2″, 4′ 7½″, 3′ 2″. **Cylinders**: 17″ x 24″ (O).
Valve Gear: Stephenson. Piston valves. **Tractive Effort**: 21 250 lbf.
Power Classification: C (4MT). **Restriction**: Yellow.

4555	"WARRIOR"	Dartmouth Steam Railway	Swindon 1924
4561		West Somerset Railway	Swindon 1924
4566		Severn Valley Railway	Swindon 1924
4588§	"TROJAN"	Dartmouth Steam Railway	Swindon 1927
5521§		Dean Forest Railway	Swindon 1927
5526§		South Devon Railway	Swindon 1928
5532§		Llangollen Railway	Swindon 1928
5538§		Flour Mill Workshop, Bream	Swindon 1928
5539§		Llangollen Railway	Swindon 1928
5541§		Dean Forest Railway	Swindon 1928
5542§		West Somerset Railway	Swindon 1928
5552§		Bodmin & Wenford Railway	Swindon 1928
5553§		West Somerset Railway	Swindon 1928
5572§		Didcot Railway Centre	Swindon 1929

4900 CLASS HALL 4-6-0

Built: 1928–43. Collett development of Churchward "Saint" Class. 259 in class. (4900 rebuilt from Saint) 4901–99, 5900–99, 6900–58 built as Halls).
Boiler Pressure: 225 lbf/sq in superheated. **Weight–Loco**: 75 tons.
Wheel Diameters: 3′ 2″, 6′ 0″. **–Tender**: 46.7 tons.
Cylinders: 18½″ x 30″ (O). **Valve Gear**: Stephenson. Piston valves.
Tractive Effort: 27 270 lbf.
Power Classification: D (5MT). **Restriction**: Red.

4920	DUMBLETON HALL	South Devon Railway	Swindon 1929
4930	HAGLEY HALL	Severn Valley Railway	Swindon 1929
4936	KINLET HALL	Tyseley Locomotive Works	Swindon 1929
4953	PITCHFORD HALL	Tyseley Locomotive Works	Swindon 1929
4965	ROOD ASHTON HALL	Tyseley Locomotive Works	Swindon 1930
4979	WOOTTON HALL	Appleby Heritage Centre	Swindon 1930
5900	HINDERTON HALL	Didcot Railway Centre	Swindon 1931
5952	COGAN HALL	Llangollen Railway	Swindon 1935
5967	BICKMARSH HALL	Northampton & Lamport Railway	Swindon 1937
5972	OLTON HALL	West Coast Railway Company, Carnforth	Swindon 1937

4965 previously carried the name 4983 ALBERT HALL.
5972 runs as "HOGWARTS CASTLE" as used in the "Harry Potter" films.

5101 CLASS 2-6-2T

Built: 1929–49. Collett development of Churchward 3100 class. 180 built (5101–99, 4100–79).
Boiler Pressure: 200 lbf/sq in superheated. **Weight:** 78.45 tons.
Wheel Diameters: 3′ 2″, 5′ 8″, 3′ 8″. **Cylinders:** 18″ x 30″ (O).
Valve Gear: Stephenson. Piston valves. **Tractive Effort:** 24300 lbf.
Power Classification: D (4MT). **Restriction:** Yellow.

4110	Tyseley Locomotive Works	Swindon 1936
4115	Barry Rail Centre	Swindon 1936
4121	Tyseley Locomotive Works	Swindon 1937
4141	Llangollen Railway	Swindon 1946
4144	Didcot Railway Centre	Swindon 1946
4150	Severn Valley Railway	Swindon 1947
4160	West Somerset Railway	Swindon 1948
5164	Severn Valley Railway	Swindon 1930
5199	Churnet Valley Railway	Swindon 1934

In addition 5193 has been rebuilt by the West Somerset Railway as 4300 Class 2-6-0 tender locomotive No. 9351. See New Build section.

5205 CLASS 2-8-0T

Built: 1923–25/40. Collett development of 4200 class. 60 built (5205–64).
Boiler Pressure: 200 lbf/sq in superheated. **Weight:** 82.1 tons.
Wheel Diameters: 3′ 2″, 4′ 7½″. **Cylinders:** 19″ x 30″ (O).
Valve Gear: Stephenson. Piston valves. **Tractive Effort:** 33170 lbf.
Power Classification: E (8F). **Restriction:** Red.

5224		Mid Hants Railway	Swindon 1925
5227		Barry Rail Centre	Swindon 1924
5239	"GOLIATH"	Dartmouth Steam Railway	Swindon 1924

5600 CLASS 0-6-2T

Built: 1924–28. Collett design. 200 built (5600–99, 6600–99).
Boiler Pressure: 200 lbf/sq in superheated. **Weight:** 68 tons.
Wheel Diameters: 4′ 7½″, 3′ 8″. **Cylinders:** 18″ x 26″ (I).
Valve Gear: Stephenson. Piston valves. **Tractive Effort:** 25800 lbf.
Power Classification: D (5MT). **Restriction:** Red.

5619	North Norfolk Railway	Swindon 1925
5637	East Somerset Railway	Swindon 1925
5643	Llangollen Railway	Swindon 1925
5668	Pontypool & Blaenavon Railway	Swindon 1926
6619	North Yorkshire Moors Railway	Swindon 1928
6634	Severn Valley Railway	Swindon 1928
6686	Barry Rail Centre	AW 974/1928
6695	Swanage Railway	AW 983/1928
6697	Didcot Railway Centre	AW 985/1928

5619 is on loan from the Telford Steam Railway.

5700 CLASS 0-6-0PT

Built: 1929–49. Collett design. The standard GWR shunter. 863 built (5700–99, 6700–79, 7700–99, 8700–99, 3700–99, 3600–99, 4600–99, 9600–82, 9700–9799). Six of the preserved examples saw use with London Transport following withdrawal by British Railways.
Boiler Pressure: 200 lbf/sq in. **Weight:** 47.5 tons (§ 49 tons).
Wheel Diameter: 4′ 7½″. **Cylinders:** 17½″ x 24″ (I).
Valve Gear: Stephenson. Slide valves. **Tractive Effort:** 22510 lbf.
Power Classification: C (4F). **Restriction:** Blue (Yellow from 1950).

GWR	*LTE*		
3650§		Didcot Railway Centre	Swindon 1939
3738§		Didcot Railway Centre	Swindon 1937
4612§		Bodmin & Wenford Railway	Swindon 1942

5764	L95	Severn Valley Railway	Swindon 1929
5775	L89	Keighley & Worth Valley Railway	Swindon 1929
5786	L92	South Devon Railway	Swindon 1930
7714		Severn Valley Railway	KS 4449/1930
7715	L99	North Norfolk Railway	KS 4450/1930
7752	L94	Tyseley Locomotive Works	NBL 24040/1930
7754		Llangollen Railway	NBL 24042/1930
7760	L90	Tyseley Locomotive Works	NBL 24048/1930
9600§		Tyseley Locomotive Works	Swindon 1945
9629§		Pontypool & Blaenavon Railway	Swindon 1946
9642§		Rye Farm, Wishaw	Swindon 1946
9681§		Dean Forest Railway	Swindon 1949
9682§		Southall Depot, London	Swindon 1949

6000 CLASS KING 4-6-0

Built: 1927–30. Collett design. 30 built.
Boiler Pressure: 250 lbf/sq in superheated. **Weight–Loco**: 89 tons.
Wheel Diameters: 3′ 0″, 6′ 6″. **–Tender**: 46.7 tons.
Cylinders: 16¼″ x 28″ (4). **Tractive Effort**: 40290 lbf.
Valve Gear: Inside Walschaerts. Piston valves. Rocking levers for outside valves.
Power Classification: Special (8P). **Restriction**: Double Red.

x–Dual (air/vacuum) brakes.

6000	KING GEORGE V	National Railway Museum, York (N)	Swindon 1927
6023	KING EDWARD II	Didcot Railway Centre	Swindon 1930
6024x	KING EDWARD I	West Somerset Railway	Swindon 1930

6100 CLASS 2-6-2T

Built: 1931–35. Collett development of 5100. 70 built (6100–69).
Boiler Pressure: 225 lbf/sq in superheated. **Weight**: 78.45 tons.
Wheel Diameters: 3′ 2″, 5′ 8″, 3′ 8″. **Cylinders**: 18″ x 30″ (O).
Valve Gear: Stephenson. Piston valves. **Tractive Effort**: 27340 lbf.
Power Classification: D (5MT). **Restriction**: Blue.

6106	Didcot Railway Centre	Swindon 1931

6400 CLASS 0-6-0PT

Built: 1932–37. Collett design. Push & Pull fitted. 40 built (6400–39).
Boiler Pressure: 165 lbf/sq in **Weight**: 45.6 tons.
Wheel Diameter: 4′ 7½″. **Cylinders**: 16½″ x 24″ (I).
Valve Gear: Stephenson. Slide valves.. **Tractive Effort**: 16510 lbf.
Power Classification: A (2P). **Restriction**: Yellow.

6412		South Devon Railway	Swindon 1934
6430		Llangollen Railway	Swindon 1937
6435	"AJAX"	Bodmin & Wenford Railway	Swindon 1937

6959 CLASS MODIFIED HALL 4-6-0

Built: 1944–49. Hawksworth development of "Hall". 71 built (6959–99, 7900–29).
Boiler Pressure: 225 lbf/sq in superheated. **Weight–Loco**: 75.8 tons.
Wheel Diameters: 3′ 2″, 6′ 0″. **–Tender**: 47.3 tons.
Cylinders: 18½″ x 30″ (O). **Valve Gear**: Stephenson. Piston valves.
Tractive Effort: 27270 lbf.
Power Classification: D (5MT). **Restriction**: Blue.

6960	RAVENINGHAM HALL	West Somerset Railway	Swindon 1944
6984	OWSDEN HALL	Gloucestershire Warwickshire Railway	Swindon 1948
6989	WIGHTWICK HALL	Buckinghamshire Railway Centre	Swindon 1948
6990	WITHERSLACK HALL	Great Central Railway	Swindon 1948
6998	BURTON AGNES HALL	Didcot Railway Centre	Swindon 1949
7903	FOREMARKE HALL	Gloucestershire Warwickshire Railway	Swindon 1949

7200 CLASS 2-8-2T

Built: 1934–50. Collett rebuilds of 4200 and 5205 class 2-8-0Ts. 54 built (7200–53).
Boiler Pressure: 200 lbf/sq in superheated. **Weight:** 92.6 tons.
Wheel Diameters: 3′ 2″, 4′ 7½″, 3′ 8″. **Cylinders:** 19″ x 30″ (O).
Valve Gear: Stephenson. Piston valves. **Tractive Effort:** 33 170 lbf
Power Classification: E (8F). **Restriction:** Blue.

7200	(rebuilt from 5277)	Buckinghamshire Railway Centre	Swindon 1930 reb. 1934
7202	(rebuilt from 5275)	Didcot Railway Centre	Swindon 1930 reb. 1934
7229	(rebuilt from 5264)	East Lancashire Railway	Swindon 1926 reb. 1935

7800 CLASS MANOR 4-6-0

Built: 1938–50. Collett design for secondary main lines. 30 built (7800–29).
Boiler Pressure: 225 lbf/sq in superheated. **Weight–Loco:** 68.9 tons.
Wheel Diameters: 3′ 0″, 5′ 8″. **–Tender:** 40 tons.
Cylinders: 18″ x 30″ (O). **Valve Gear:** Stephenson. Piston valves.
Tractive Effort: 27 340 lbf.
Power Classification: D (5MT). **Restriction:** Blue.

7802	BRADLEY MANOR	Severn Valley Railway	Swindon 1938
7808	COOKHAM MANOR	Didcot Railway Centre	Swindon 1938
7812	ERLESTOKE MANOR	Severn Valley Railway	Swindon 1939
7819	HINTON MANOR	Designer Outlet Village, Swindon	Swindon 1939
7820	DINMORE MANOR	Tyseley Locomotive Works	Swindon 1950
7821	DITCHEAT MANOR	Steam – Museum of the Great Western Railway (N)	Swindon 1950
7822	FOXCOTE MANOR	Llangollen Railway	Swindon 1950
7827	LYDHAM MANOR	Dartmouth Steam Railway	Swindon 1950
7828	ODNEY MANOR	West Somerset Railway	Swindon 1950

9400 CLASS 0-6-0PT

Built: 1947–56. Hawksworth design. 210 built (9400–99, 8400–99, 3400–09).
Boiler Pressure: 200 lbf/sq in (* Superheated) **Weight:** 55.35 tons.
Wheel Diameter: 4′ 7½″. **Cylinders:** 17½″ x 24″ (I).
Valve Gear: Stephenson. Slide valves **Tractive Effort:** 22 510 lbf.
Power Classification: C (4F). **Restriction:** Red.

9400*	Steam – Museum of the Great Western Railway (N)	Swindon 1947
9466	Barry Rail Centre	RSH 7617/1952

BURRY PORT & GWENDRAETH VALLEY RAILWAY 0-6-0ST

Built: 1900. **Weight:** 29 tons.
Wheel Diameter: 3′ 6″. **Cylinders:** 14″ x 20″(O)
This loco was supplied new to the BPGVR and was sold into industrial service in 1914.

2	PONTYBEREM	Didcot Railway Centre	AE 1421/1900

SANDY & POTTON RAILWAY 0-4-0WT

Built: 1857. The Sandy & Potton Railway became part of the LNWR and the loco worked on the Cromford & High Peak Railway from 1863–1878. The loco was sold to the Wantage Tramway in 1878.
Boiler Pressure: 120 lbf/sq in. **Weight:** 15 tons.
Wheel Diameter: 3′ 0″ **Cylinders:** 9″ x 12″ (O)
Tractive Effort: 5510 lbf.

SPR	*LNWR*	*WT*		
SHANNON	1863	5	Didcot Railway Centre (N)	GE 1857

SOUTH DEVON RAILWAY 0-4-0WT

Built: 1868. Vertical boilered locomotive. **Gauge:** 7′ 0¼″.
Wheel Diameter: 3′ 0″. **Cylinders:** 9″ x 12″ (V).

GWR	*SDR*		
2180	151 TINY	South Devon Railway (N)	Sara 1868

▲ Emerging in BR Blue after a lengthy restoration from Barry condition, King Class 4-6-0 6023 "KING EDWARD II" basks in the sunshine at Didcot Railway Centre on 29 May 2011. **Brian Garvin**

▼ Manor Class 4-6-0 7827 "LYDHAM MANOR" has been one of the regular performers at the Dartmouth Steam Railway for many years. On 18 April 2011 it crosses Broadsands Viaduct with the 10.30 Paignton–Kingswear. **Glen Batten**

SOUTHERN

1.2. SOUTHERN RAILWAY AND CONSTITUENT COMPANIES' STEAM LOCOMOTIVES

GENERAL

The Southern Railway (SR) was an amalgamation of the London, Brighton & South Coast Railway (LBSCR), the London & South Western Railway (LSWR) and the South Eastern & Chatham Railway (SECR). The last of these was formed in 1898 by the amalgamation of the South Eastern Railway (SER) and London, Chatham & Dover Railway (LCDR).

LOCOMOTIVE NUMBERING SYSTEM

On formation of the SR in 1924, all locomotives (including new builds) were given a prefix letter to denote the works which maintained them as follows:

A Ashford Works. All former SECR locomotives plus some D1, L1, U1.
B Brighton Works. All former LBSCR locomotives plus some D1.
E Eastleigh Works. All former LSWR locomotives plus LN, V, Z.

In 1931 locomotives were renumbered. "E" prefix locomotives merely lost the prefix (except locomotives with an "0" in front of the number to which 3000 was added). "A" prefix locomotives had 1000 added and "B" prefix locomotives had 2000 added, e.g. B636 became 2636, E0298 became 3298.

In 1941 Bulleid developed a most curious numbering system for his new locomotives. This consisted of two numbers representing the numbers of leading and trailing axles respectively followed by a letter denoting the number of the driving axles. This was followed by the locomotive serial number. The first pacific was therefore 21C1, and the first Q1 0-6-0 was C1.

In 1948 British Railways added 30000 to all numbers, but the 3xxx series (formerly 0xxx series) were totally renumbered. The Q1s became 33xxx, the MNs 35xxx and the WCs 34xxx. Isle of Wight locomotives had their own number series, denoted by a "W" prefix. This indicated Ryde Works maintenance and was carried until the end of steam.

In the section which follows, locomotives are listed generally in order of BR numbers. Three old locomotives which were withdrawn before nationalisation are listed at the end of the section.

CLASSIFICATION

The LBSCR originally classified locomotive classes by a letter which denoted the use of the class. A further development was to add a number, to identify different classes of similar use. A rebuild was signified by an "X" suffix. In its latter years, new classes of different wheel arrangement were given different letters. The SECR gave each class a letter. A number after the letter signified either a new class which was a modification of the original or a rebuild. The SR perpetuated this system. The LSWR had an odd system based on the works order number for the first locomotive of the class to be built. These went A1, B1......Z1, A2......Z2, A3......etc. and did not only apply to locomotives. Locomotives bought from outside contractors were classified by the first number to be delivered, e.g. "0298 Class".

CLASS 02 — 0-4-4T

Built: 1889–91. Adams LSWR design.
Boiler Pressure: 160 lbf/sq in.
Wheel Diameters: 4′ 10″, 3′ 1″.
Valve Gear: Stephenson. Slide valves.
BR Power Classification: 1P.
Weight: 48.4 tons.
Cylinders: 17″ x 24″ (I).
Tractive Effort: 17235 lbf.
Air braked.

BR	*SR*	*LSWR*		
W24	E209–W24	209 CALBOURNE	Isle of Wight Steam Railway	Nine Elms 341/1891

CLASS M7 — 0-4-4T

Built: 1897–1911. Drummond LSWR design. 105 built.
Boiler Pressure: 175 lbf/sq in.
Wheel Diameters: 5′ 7″, 3′ 7″.
Valve Gear: Stephenson. Slide valves.
BR Power Classification: 2P.
Weight: 60.15 tons.
Cylinders: 18½″ x 26″ (I).
Tractive Effort: 19760 lbf.

30053 was push-pull fitted and air braked.

BR	*SR*	*LSWR*		
30053	E53–53	53	Swanage Railway	Nine Elms 1905
30245	E245–245	245	National Railway Museum, York (N)	Nine Elms 501/1897

CLASS USA — 0-6-0T

Built: 1942–43 by Vulcan Works, Wilkes-Barre, PA, USA for US Army Transportation Corps. 93 built (a further 289 were built by other builders). 15 were sold to SR in 1947 of which 14 became 61–74.
Boiler Pressure: 210 lbf/sq in.
Wheel Diameter: 4′ 6″.
Valve Gear: Walschaerts. Piston valves.
BR Power Classification: 3F.
Weight: 46.5 tons.
Cylinders: 16½″ x 24″ (O).
Tractive Effort: 21600 lbf.

BR	*SR*	*USATC*	*Present*		
30064	64	1959		Bluebell Railway	VIW 4432/1943
30065	65	1968	MAUNSELL	Kent & East Sussex Railway	VIW 4441/1943
30070	70	1960	WAINWRIGHT	Kent & East Sussex Railway	VIW 4433/1943
30072	72	1973		Keighley & Worth Valley Railway	VIW 4446/1943

30065 carried DS237 and 30070 carried DS238 from 1963.

CLASS B4 — 0-4-0T

Built: 1891–1909. Adams LSWR design for dock shunting. 25 built.
Boiler Pressure: 140 lbf/sq in.
Wheel Diameter: 3′ 9¾″.
Valve Gear: Stephenson. Slide valves.
BR Power Classification: 1F.
Weight: 33.45 tons.
Cylinders: 16″ x 22″ (O).
Tractive Effort: 14650 lbf.

BR	*SR*			
30096	E96–96	NORMANDY	Bluebell Railway	Nine Elms 396/1893
30102	E102–102	GRANVILLE	Bressingham Steam Museum	Nine Elms 406/1893

CLASS T9 — 4-4-0

Built: 1889–1924. Drummond LSWR express passenger design. 66 built.
Boiler Pressure: 175 lbf/sq in superheated
Wheel Diameters: 3′ 7″, 6′ 7″.
Cylinders: 19″ x 26″ (I).
Tractive Effort: 17670 lbf.
Weight–Loco: 51.8 tons.
–Tender: 44.85 tons.
Valve Gear: Stephenson. Slide valves.
BR Power Classification: 3P.

BR	*SR*	*LSWR*		
30120	E120–120	120	Bodmin & Wenford Railway (N)	Nine Elms 572/1899

▲ Former L&SWR 0298 Class 2-4-0WT 30585 stands at Quainton Road station at the Buckinghamshire Railway Centre on 7 October 2006. **Courtesy Buckinghamshire Railway Centre**

▼ Lord Nelson Class 850 "LORD NELSON" climbs through Chawton Park Woods with a demonstration freight from Alton to Alresford during the Mid Hants Railway spring gala on 25 March 2011. **Jon Bowers**

CLASS S15 (URIE) 4-6-0

Built: 1920–21. Urie LSWR design. 20 built.
Boiler Pressure: 180 lbf/sq in superheated.
Wheel Diameters: 3′ 7″, 5′ 7″.
Cylinders: 21″ x 28″ (O).
Tractive Effort: 28200 lbf.
Weight–Loco: 79.8 tons.
–Tender: 57.8 tons.
Valve Gear: Walschaerts. Piston valves.
BR Power Classification: 6F.

BR	*SR*	*LSWR*		
30499	E499–499	499	Mid Hants Railway	Eastleigh 1920
30506	E506–506	506	Mid Hants Railway	Eastleigh 1920

CLASS Q 0-6-0

Built: 1938–39. Maunsell SR design. 20 built (30530–49).
Boiler Pressure: 200 lbf/sq in superheated.
Wheel Diameter: 5′ 1″.
Cylinders: 19″ x 26″ (I).
Tractive Effort: 26160 lbf.
Weight–Loco: 49.5 tons.
–Tender: 40.5 tons.
Valve Gear: Stephenson. Piston valves.
BR Power Classification: 4F.

BR	*SR*		
30541	541	Bluebell Railway	Eastleigh 1939

0415 CLASS 4-4-2T

Built: 1882–85. Adams LSWR design. 72 built.
Boiler Pressure: 160 lbf/sq in.
Wheel Diameters: 3′ 0″, 5′ 7″, 3′ 0″
Valve Gear: Stephenson. Slide valves.
BR Power Classification: 1P.
Weight: 55.25 tons.
Cylinders: 17½″ x 24″ (O).
Tractive Effort: 14920 lbf.

BR	*SR*	*LSWR*		
30583	E0488–3488	488	Bluebell Railway	N 3209/1885

0298 CLASS 2-4-0WT

Built: 1863–75. WG Beattie LSWR design. Last used on the Wenfordbridge branch in Cornwall. 85 built. Survivors reboilered in 1921.
Boiler Pressure: 160 lbf/sq in.
Wheel Diameters: 3′ 7¾″, 5′ 7″.
Valve Gear: Allan.
BR Power Classification: 0P.
Weight: 35.75 (§ 36.3) tons.
Cylinders: 16½″ x 22″ (O).
Tractive Effort: 12160 lbf.

BR	*SR*	*LSWR*		
30585§	E0314–3314	0314	Buckinghamshire Railway Centre	BP 1414/1874 reb Elh 1921
30587	E0298–3298	0298	Bodmin & Wenford Railway (N)	BP 1412/1874 reb Elh 1921

CLASS N15 KING ARTHUR 4-6-0

Built: 1925–27. Maunsell SR development of Urie LSWR design. 54 built (30448–457, 30763–806).
Boiler Pressure: 200 lbf/sq in superheated.
Wheel Diameters: 3′ 7″, 6′ 7″.
Cylinders: 20½″ x 28″ (O).
Tractive Effort: 25320 lbf.
Weight–Loco: 80.7 tons.
–Tender: 57.5 tons.
Valve Gear: Walschaerts. Piston valves.
BR Power Classification: 5P.

BR	*SR*			
30777	777	SIR LAMIEL	Great Central Railway (N)	NBL 23223/1925

CLASS S15 (MAUNSELL) 4-6-0

Built: 1927–36. Maunsell SR development of Urie LSWR design.
Boiler Pressure: 200 lbf/sq in superheated.
Wheel Diameters: 3′ 7″, 5′ 7″.
Cylinders: 20½″ x 28″ (O).
Tractive Effort: 29860 lbf.
Weight–Loco: 80.7 (* 79.25) tons.
–Tender: 56.4 tons.
Valve Gear: Walschaerts. Piston valves.
BR Power Classification: 6F.

BR	*SR*			
30825	825		North Yorkshire Moors Railway	Eastleigh 1927
30828	828	"HARRY A FRITH"	Mid Hants Railway	Eastleigh 1927
30830	830		North Yorkshire Moors Railway	Eastleigh 1927
30847*	847		Bluebell Railway	Eastleigh 1936

30825 has been restored using a substantial number of components from 30841.

CLASS LN LORD NELSON 4-6-0

Built: 1926–29. Maunsell SR design. 16 built (30850–65).
Boiler Pressure: 220 lbf/sq in superheated.
Wheel Diameters: 3′ 1″, 6′ 7″.
Cylinders: 16½″ x 26″ (4).
Tractive Effort: 33510 lbf.
Weight–Loco: 83.5 tons.
–Tender: 57.95 tons.
Valve Gear: Walschaerts. Piston valves.
BR Power Classification: 7P.

Dual (air/vacuum) brakes.

BR	*SR*			
30850	E850–850	LORD NELSON	Mid Hants Railway (N)	Eastleigh 1926

CLASS V SCHOOLS 4-4-0

Built: 1930–35. Maunsell SR design. 40 built (30900–39).
Boiler Pressure: 220 lbf/sq in superheated.
Wheel Diameters: 3′ 1″, 6′ 7″.
Cylinders: 16½″ x 26″ (3).
Tractive Effort: 25130 lbf.
Weight–Loco: 67.1 tons.
–Tender: 42.4 tons.
Valve Gear: Walschaerts. Piston valves.
BR Power Classification: 5P.

BR	*SR*			
30925	925	CHELTENHAM	National Railway Museum, York (N)	Eastleigh 1934
30926	926	REPTON	North Yorkshire Moors Railway	Eastleigh 1934
30928	928	STOWE	Bluebell Railway	Eastleigh 1934

CLASS P 0-6-0T

Built: 1909–10. Wainwright SECR design. 8 built.
Boiler Pressure: 160 lbf/sq in.
Wheel Diameter: 3′ 9″.
Valve Gear: Stephenson. Slide valves.
BR Power Classification: 0F.
Weight: 28.5 tons.
Cylinders: 12″ x 18″ (I).
Tractive Effort: 7830 lbf.

BR	*SR*	*SECR*		
31027	A 27–1027	27	Bluebell Railway	Ashford 1910
31178	A178–1178	178 "PIONEER II"	Bluebell Railway	Ashford 1910
31323	A323–1323	323	Bluebell Railway	Ashford 1910
31556	556–A556–1556	753–5753	Kent & East Sussex Railway	Ashford 1909

CLASS 01 0-6-0

Built: 1903–15. Wainwright SECR design. 66 built. 59 were rebuilt out of 122 "O" class.
Boiler Pressure: 150 lbf/sq in.
Wheel Diameter: 5′ 1″.
Cylinders: 18″ x 26″ (I).
Tractive Effort: 17610 lbf.
Weight–Loco: 41.05 tons.
–Tender: 28.20 tons.
Valve Gear: Stephenson. Slide valves.
BR Power Classification: 1F.

BR	*SR*	*SECR*		
31065	A65–1065	65	Bluebell Railway	Ashford 1896 reb. 1908

▲ Schools Class 4-4-0 30926 "REPTON" is one of the North Yorkshire Moors Railway locos cleared to operate to Whitby. On 4 October 2009 it passes Sleights with a Whitby–Pickering train.
Andrew Mason

▼ P Class 0-6-0T 323 returned to traffic on the Bluebell Railway in spring 2011. It takes water at Sheffield Park between duties on 12 March 2011. **Phil Barnes**

▲ U Class 31806 arrives at Medstead & Four Marks on the Mid Hants Railway with the 11.50 Alresford–Alton on 26 March 2011. **Phil Barnes**

▼ E4 Class 0-6-2T 473 "BIRCH GROVE" heads an early morning train for Sheffield Park away from Horsted Keynes on 22 October 2010 during the Bluebell Railway's autumn 2010 gala weekend. **Jon Bowers**

CLASS H — 0-4-4T

Built: 1904–15. Wainwright SECR design. 66 built.
Boiler Pressure: 160 lbf/sq in. **Weight:** 54.4 tons.
Wheel Diameters: 5′ 6″, 3′ 7″. **Cylinders:** 18″ x 26″ (I).
Valve Gear: Stephenson. Slide valves. **Tractive Effort:** 17 360 lbf.
BR Power Classification: 1P.

Push and Pull fitted. Dual (air/vacuum) brakes..

BR	*SR*	*SECR*		
31263	A263–1263	263	Bluebell Railway	Ashford 1905

CLASS C — 0-6-0

Built: 1900–08. Wainwright SECR design. 109 built.
Boiler Pressure: 160 lbf/sq in. **Weight–Loco:** 43.8 tons.
Wheel Diameter: 5′ 2″. **–Tender:** 38.25 tons.
Cylinders: 18½″ x 26″ (I). **Valve Gear:** Stephenson. Slide valves.
Tractive Effort: 19 520 lbf. **BR Power Classification:** 2F.

BR	*SR*	*SECR*		
31592–DS239	A592–1592	592	Bluebell Railway	Longhedge 1902

CLASS U — 2-6-0

Built: 1928–31. Maunsell SR design. 50 built (31610–39, 31790–809). 31790–809 were converted from class K (River Class) 2-6-4Ts.
Boiler Pressure: 200 lbf/sq in superheated. **Weight–Loco:** 61.9 (* 62.55) tons.
Wheel Diameters: 3′ 1″, 6′ 0″. **–Tender:** 42.4 tons.
Cylinders: 19″ x 28″ (O). **Valve Gear:** Walschaerts. Piston valves.
Tractive Effort: 23 870 lbf. **BR Power Classification:** 4MT.

* Formerly Class K 2–6–4T A806 RIVER TORRIDGE built Ashford 1926.

BR	*SR*		
31618	A618–1618	Bluebell Railway	Brighton 1928
31625	A625–1625	Mid Hants Railway	Ashford 1929
31638	A638–1638	Bluebell Railway	Ashford 1931
31806*	A806–1806	Mid Hants Railway	Brighton 1928

CLASS D — 4-4-0

Built: 1901–07. Wainwright SECR design. 51 built.
Boiler Pressure: 175 lbf/sq in. **Weight–Loco:** 50 tons.
Wheel Diameters: 3′ 7″, 6′ 8″. **–Tender:** 39.1 tons.
Cylinders: 19¼″ x 26″ (I). **Valve Gear:** Stephenson. Slide valves.
Tractive Effort: 17 910 lbf. **BR Power Classification:** 1P.

BR	*SR*	*SECR*		
31737	A737–1737	737	National Railway Museum, York (N)	Ashford 1901

CLASS N — 2-6-0

Built: 1917–34. Maunsell SECR design. Some built by SR. 80 built.
Boiler Pressure: 200 lbf/sq in. **Weight–Loco:** 59.4 tons.
Wheel Diameters: 3′ 1″, 5′ 6″. **–Tender:** 39.25 tons.
Cylinders: 19″ x 28″ (O). **Valve Gear:** Walschaerts. Piston valves.
Tractive Effort: 26 040 lbf. **BR Power Classification:** 4MT.

BR	*SR*	*Present*		
31874	A874–1874	"5 JAMES"	Mid Hants Railway	Woolwich Arsenal 1925

CLASS E1 — 0-6-0T

Built: 1874–83. Stroudley LBSCR design. 80 built.
Boiler Pressure: 160 lbf/sq in.
Weight: 44.15 tons.
Wheel Diameter: 4′ 6″.
Cylinders: 17″ x 24″ (I).
Valve Gear: Stephenson. Slide valves.
Tractive Effort: 17470 lbf.

BR	*SR*	*LBSCR*			
–	B110	110	BURGUNDY	East Somerset Railway	Brighton 1877

CLASS E4 — 0-6-2T

Built: 1897–1903. R Billinton LBSCR design. 120 built.
Boiler Pressure: 160 lbf/sq in.
Weight: 56.75 tons.
Wheel Diameters: 5′ 0″, 4′ 0″.
Cylinders: 18″ x 26″ (I).
Valve Gear: Stephenson. Slide valves.
Tractive Effort: 19090 lbf.
BR Power Classification: 2MT.

BR	*SR*	*LBSCR*			
32473	B473–2473	473	BIRCH GROVE	Bluebell Railway	Brighton 1898

CLASSES A1 & A1X — "TERRIER" — 0-6-0T

Built: 1872–80 as Class A1*. Stroudley LBSCR design. Most rebuilt to A1X from 1911. 50 built.
Boiler Pressure: 150 lbf/sq in.
Weight: 28.25 tons.
Wheel Diameter: 4′ 0″.
Cylinders: 14″ († 13″, * 12″) x 20″ (I).
Valve Gear: Stephenson. Slide valves.
Tractive Effort: 10410 lbf († 8890 lbf, * 7650 lbf).
BR Power Classification: 0P.

a Air brakes. x Dual (air/vacuum) brakes.

BR	*SR*	*LBSCR*			
32636 x†	B636–2636	72	FENCHURCH	Bluebell Railway	Brighton 1872
32640 a	W11–2640	40	NEWPORT	Isle of Wight Steam Railway	Brighton 1878
32646 a	W2–W8	46–646	FRESHWATER	Isle of Wight Steam Railway	Brighton 1876
32650 x*	B650–W9	50–650	WHITECHAPEL	Spa Valley Railway	Brighton 1876
DS680 a*	A751–680S	54–654	WADDON	Canadian Railroad Historical Museum	Brighton 1875
32655	B655–2655	55–655	STEPNEY	Bluebell Railway	Brighton 1875
32662 a*	B662–2662	62–662	MARTELLO	Bressingham Steam Museum	Brighton 1875
32670		70	POPLAR	Kent & East Sussex Railway	Brighton 1872
32678 x	B678–W4–W14	78–678	KNOWLE	Kent & East Sussex Railway	Brighton 1880
– a	380S	82–682	BOXHILL	National Railway Museum, York (N)	Brighton 1880

32640 was also named BRIGHTON and was originally Isle of Wight Central Railway No. 11.
32646 was originally Freshwater Yarmouth and Newport Railway No. 2. It was sold to the LSWR and became 734. It has also been named NEWINGTON.
32650 became 515S (departmental) and was named FISHBOURNE when on the Isle of Wight. It is now named "SUTTON".
DS680 was sold to the SECR and became their 75.
32678 was named BEMBRIDGE when on the Isle of Wight.

CLASS Q1 — 0-6-0

Built: 1942. Bulleid SR "Austerity" design. 40 built (33001–40).
Boiler Pressure: 230 lbf/sq in superheated.
Weight–Loco: 51.25 tons.
Wheel Diameter: 5′ 1″.
–Tender: 38 tons.
Cylinders: 19″ x 26″ (I).
Valve Gear: Stephenson. Slide valves.
Tractive Effort: 30080 lbf
BR Power Classification: 5F.

BR	*SR*		
33001	C1	National Railway Museum, York (N)	Brighton 1942

▲ Two A1X "Terriers", 32678 and 32670, top-and-tail one coach as the 10.50 Bodiam–Tenterden at the bottom of Tenterden Bank on 2 May 2011. **Phil Barnes**

▼ Rebuilt West Country 4-6-2 34028 "EDDYSTONE" leaves Swanage with the 16.00 to Norden on 23 April 2011. **Chris Wilson**

CLASSES WC & BB WEST COUNTRY and BATTLE OF BRITAIN 4-6-2

Built: 1945–51. Bulleid SR design with "air smoothed" casing and Boxpox driving wheels. 110 built (34001–110). All preserved examples built at Brighton except 34101 (Eastleigh).
Boiler Pressure: 250 lbf/sq in superheated. **Weight–Loco:** 86 (* 91.65) tons.
Wheel Diameters: 3′ 1″, 6′ 2″, 3′ 1″ **–Tender:** 42.7, 47.9 or 47.75 tons.
Cylinders: $16^3/_8$″ x 24″ (3).
Valve Gear: Bulleid chain driven (* Walschaerts. Piston valves).
Tractive Effort: 27720 lbf. **BR Power Classification:** 7P.

* Rebuilt at Eastleigh by Jarvis 1957–61 with the removal of the air-smoothed casing.
x Dual (air/vacuum) brakes.

BR	*SR*			
34007	21C107	WADEBRIDGE	Mid Hants Railway	1945
34010*	21C110	SIDMOUTH	Swanage Railway	1945 reb 1959
34016*	21C116	BODMIN	Mid Hants Railway	1945 reb 1958
34023	21C123	BLACKMORE VALE	Bluebell Railway	1946
34027*x	21C127	TAW VALLEY	Severn Valley Railway	1946 reb 1957
34028*	21C128	EDDYSTONE	Swanage Railway	1946 reb 1958
34039*	21C139	BOSCASTLE	Great Central Railway	1946 reb 1959
34046*x	21C146	BRAUNTON	West Somerset Railway	1946 reb 1959
34051	21C151	WINSTON CHURCHILL	National Railway Museum, York (N)	1946
34053*	21C153	SIR KEITH PARK	Swanage Railway	1947 reb 1958
34058*	21C158	SIR FREDERICK PILE	Avon Valley Railway	1947 reb 1960
34059*	21C159	SIR ARCHIBALD SINCLAIR	Bluebell Railway	1947 reb 1960
34067 x	21C167	TANGMERE	Southall Depot, London	1947
34070	21C170	MANSTON	Swanage Railway	1947
34072		257 SQUADRON	Swanage Railway	1948
34073		249 SQUADRON	East Lancashire Railway	1948
34081		92 SQUADRON	Nene Valley Railway	1948
34092		CITY OF WELLS	Keighley & Worth Valley Railway	1949
34101*		HARTLAND	North Yorkshire Moors Railway	1950 reb 1960
34105		SWANAGE	Mid Hants Railway	1950

34010 will be restored as "34109 SIR TRAFFORD LEIGH-MALLORY".
34023 was named BLACKMOOR VALE to 4/50.
34092 was named WELLS to 3/50.

CLASS MN MERCHANT NAVY 4-6-2

Built: 1941–49. Bulleid SR design with air smoothed casing and similar features to "WC" and "BB". All rebuilt 1956–59 by Jarvis to more conventional appearance. 30 built (35001–30). All locomotives were built and rebuilt at Eastleigh.
Boiler Pressure: 250 lbf/sq in superheated. **Weight–Loco:** 97.9 tons.
Wheel Diameters: 3′ 1″, 6′ 2″, 3′ 7″. **–Tender:** 47.8 tons.
Cylinders: 18″ x 24″ (3). **Valve Gear:** Walschaerts. Piston valves.
Tractive Effort: 33490 lbf. **BR Power Classification:** 8P.

§ Sectioned.
x Dual (air/vacuum) brakes.

BR	*SR*			
35005	21C5	CANADIAN PACIFIC	Mid Hants Railway	1941 reb 1959
35006	21C6	PENINSULAR & ORIENTAL S.N. Co.	Gloucestershire Warwickshire Railway	1941 reb 1959
35009	21C9	SHAW SAVILL	East Lancashire Railway	1942 reb 1957
35010	21C10	BLUE STAR	Colne Valley Railway	1942 reb 1957
35011	21C11	GENERAL STEAM NAVIGATION	Hope Farm, Sellindge	1944 reb 1959
35018	21C18	BRITISH INDIA LINE	GCE & SCS, Easton, Isle of Portland	1945 reb 1956
35022		HOLLAND-AMERICA LINE	Southall Depot, London	1948 reb 1956
35025		BROCKLEBANK LINE	Hope Farm, Sellindge	1948 reb 1956
35027		PORT LINE	East Lancashire Railway	1948 reb 1957
35028x		CLAN LINE	Stewarts Lane Depot, London	1948 reb 1959
35029§		ELLERMAN LINES	National Railway Museum, York (N)	1949 reb 1959

CLASS T3 — 4-4-0

Built: 1882–93. Adams LSWR design. 20 built
Boiler Pressure: 175 lbf/sq in.
Wheel Diameters: 3′ 7″, 6′ 7″.
Cylinders: 19″ x 26″ (O).
Tractive Effort: 17670 lbf.
Weight–Loco: 48.55 tons.
–Tender: 33.2 tons.
Valve Gear: Stephenson. Slide valves.

SR	*LSWR*		
E563–563	563	National Railway Museum, Shildon (N)	Nine Elms 380/1893

CLASS B1 — "GLADSTONE" — 0-4-2

Built: 1882–91. Stroudley LBSCR design. 49 built.
Boiler Pressure: 150 lbf/sq in.
Wheel Diameters: 6′ 6″, 4′ 6″.
Cylinders: 18¼″ x 26″ (I).
Tractive Effort: 14 160 lbf.
Air brakes.
Weight–Loco: 38.7 tons.
–Tender: 29.35 tons.
Valve Gear: Stephenson. Slide valves.

SR	*LBSCR*			
B618	214–618	GLADSTONE	National Railway Museum, York (N)	Brighton 1882

CANTERBURY & WHITSTABLE RAILWAY — 0-4-0

Built: 1830. Robert Stephenson & Company design.
Boiler Pressure: 40 lbf/sq in.
Wheel Diameter: 4′ 0″.
Cylinders: 10½″ x 18″ (O).
Weight–Loco: 6.25 tons
–Tender:
Tractive Effort: 2680 lbf.

INVICTA	Museum of Canterbury	RS 24/1830

▲ Unrebuilt West Country Pacific 4-6-2 34070 "MANSTON" has just left Harmans Cross with the 17.10 Swanage–Norden on 11 September 2009. **Hugh Ballantyne**

1.3. LONDON MIDLAND & SCOTTISH RAILWAY & CONSTITUENT COMPANIES' STEAM LOCOMOTIVES

GENERAL

The LMS was formed in 1923 by the amalgamation of the Midland Railway (MR), London & North Western Railway (LNWR), Caledonian Railway (CR), Glasgow & South Western Railway (GSWR) and Highland Railway (HR), plus a few smaller railways. Prior to this the North London Railway (NLR) and the Lancashire & Yorkshire Railway (L&Y) had been absorbed by the LNWR and the London, Tilbury & Southend Railway (LTSR) had been absorbed by the Midland Railway.

NUMBERING SYSTEM

Originally number series were allocated to divisions as follows:

1– 4999	Midland Division (Midland and North Staffordshire Railway).
5000– 9999	Western Division "A" (LNWR).
10000–13999	Western Division "B" (L&Y).
14000–17999	Northern Division (Scottish Railways).

From 1934 onwards, all LMS standard locomotives and new builds were numbered in the range from 1–9999, and any locomotives which would have had their numbers duplicated had 20000 added to their original number.

At nationalisation 40000 was added to all LMS numbers except that locomotives which were renumbered in the 2xxxx series were further renumbered generally in the 58xxx series. In the following section locomotives are listed in order of BR number or in the position of the BR number they would have carried if they had lasted into BR days, except for very old locomotives which are listed at the end of the section.

CLASSIFICATION SYSTEM

LMS locomotives did not generally have unique class designations but were referred to by their power classification which varied from 0 to 8 followed by the letters "P" for a passenger locomotive and "F" for a freight locomotive. BR adopted the LMS system and used the description "MT" to denote mixed traffic locos. The power classifications are generally shown in the class headings.

MIDLAND 115 CLASS (1P) "SPINNER" 4-2-2

Built: 1887–1900. Johnson design. 95 built.
Boiler Pressure: 170 lbf/sq in.
Wheel Diameters: 3′ 10″, 7′ 9½″, 4′ 4½″
Cylinders: 19″ x 26″ (I).
Tractive Effort: 15280 lbf.
Weight–Loco: 43.95 tons.
–Tender: 21.55 tons.
Valve Gear: Stephenson. Slide valves.

LMS	*MR*		
673	118–673	National Railway Museum, York (N)	Derby 1897

CLASS 4P COMPOUND 4-4-0

Built: 1902–03. Johnson Midland design, rebuilt by Deeley 1914–19 to a similar design to the Deeley compounds which were built 1905–09. A further similar batch was built by the LMS in 1924–32. 240 built (41000–41199, 40900–40939).
Boiler Pressure: 200 lbf/sq in superheated.
Wheel Diameters: 3′ 6½″, 7′ 0″.
Weight–Loco: 61.7 tons.
–Tender: 45.9 tons.
Cylinders: One high pressure. 19″ x 26″ (I).
Two low pressure. 21″ x 26″ (O).
Valve Gear: Stephenson. Slide valves on low pressure cylinders, piston valves on high pressure cylinder.
Tractive Effort: 23205 lbf.

BR	*LMS*	*MR*		
41000	1000	1000 (2631 pre-1907)	Bo'ness & Kinneil Railway (N)	Derby 1902 reb 1914

CLASS 2MT 2-6-2T

Built: 1946–52. Ivatt LMS design. 130 built (41200–41329).
Boiler Pressure: 200 lbf/sq in superheated.
Wheel Diameters: 3′ 0″, 5′ 0″, 3′ 0″.
Weight: 65.2 (* 63.25) tons.
Cylinders: 16½″ (* 16″) x 24″ (O).
Valve Gear: Walschaerts. Piston valves.
Tractive Effort: 18510 (* 17410) lbf.

41241*	Keighley & Worth Valley Railway	Crewe 1949
41298	Isle of Wight Steam Railway	Crewe 1951
41312	Mid Hants Railway	Crewe 1952
41313	Isle of Wight Steam Railway	Crewe 1952

CLASS 1F 0-6-0T

Built: 1878–99. Johnson Midland design. Rebuilt with Belpaire boiler 1926. 240 built.
Boiler Pressure: 150 lbf/sq in.
Wheel Diameter: 4′ 7″.
Valve Gear: Stephenson. Slide valves.
Weight: 45.45 tons.
Cylinders: 17″ x 24″ (I).
Tractive Effort: 16080 lbf.

BR	*LMS*		*MR*		
41708	1708	1708	1418	Barrow Hill Roundhouse	Derby 188

LTS 79 CLASS (3P) 4-4-2T

Built: 1909. Whitelegg LTSR design. 4 built.
Boiler Pressure: 170 lbf/sq in.
Wheel Diameters: 3′, 6″, 6′ 6″, 3′ 6″.
Valve Gear: Stephenson. Slide valves.
Weight: 69.35 tons.
Cylinders: 19″ x 26″ (O).
Tractive Effort: 17390 lbf.

BR	*LMS*	*MR*	*LTSR*			
41966	2148	2177	80	THUNDERSLEY	Bressingham Steam Museum (N)	RS 3367/1909

▲ Ivatt 2MT 2-6-2T 41241 climbs the gradient out of Keighley with the 11.45 to Oxenhope on 7 March 2010. **Brian Dobbs**

▼ 5MT 2-6-0 42968, the only one of this class to be preserved, leaves Highley with the 15.40 Bridgnorth–Kidderminster on 30 April 2011. **Phil Barnes**

CLASS 4MT 2-6-4T

Built: 1945–51. Fairburn modification of Stanier design (built 1936–43). This in turn was a development of a Fowler design built 1927–34. 383 built (Stanier & Fairburn) (42050–299/425–494/537–699).
Boiler Pressure: 200 lbf/sq in superheated. **Weight**: 85.25 tons.
Wheel Diameters: 3′, 3½″, 5′ 9″, 3′ 3½″. **Cylinders**: 19¾″ x 26″ (O).
Valve Gear: Walschaerts. Piston valves. **Tractive Effort**: 24670 lbf.

42073	Lakeside & Haverthwaite Railway	Brighton 1950
42085	Lakeside & Haverthwaite Railway	Brighton 1951

CLASS 4MT 2-6-4T

Built: 1934. Stanier LMS 3-cylinder design for LTSR line. 37 built (42500–36).
Boiler Pressure: 200 lbf/sq in superheated. **Weight**: 92.5 tons.
Wheel Diameters: 3′, 3½″, 5′ 9″, 3′ 3½″. **Cylinders**: 16″ x 26″ (3).
Valve Gear: Walschaerts. Piston valves. **Tractive Effort**: 24600 lbf.

BR	*LMS*		
42500	2500	National Railway Museum, York (N)	Derby 1934

CLASS 5MT "CRAB" 2-6-0

Built: 1926–32. Hughes LMS design. 245 built (42700–944).
Boiler Pressure: 180 lbf/sq in superheated. **Weight–Loco**: 66 tons.
Wheel Diameters: 3′ 6½″, 5′ 6″. **–Tender**: 42.2 (41.5*) tons.
Cylinders: 21″ x 26″ (O). **Valve Gear**: Walschaerts. Piston valves.
Tractive Effort: 26580 lbf.

BR	*LMS*		
42700	13000–2700	National Railway Museum, Shildon (N)	Horwich 1926
42765*	13065–2765	East Lancashire Railway	Crewe 5757/1927
42859	13159–2859	Binbrook Trading Estate, Lincolnshire	Crewe 5981/1930

CLASS 5MT 2-6-0

Built: 1933–34. Stanier LMS design. 40 built (42945–84).
Boiler Pressure: 225 lbf/sq in superheated. **Weight–Loco**: 69.1 tons.
Wheel Diameters: 3′ 3½″, 5′ 6″. **–Tender**: 42.2 tons.
Cylinders: 18″ x 28″ (O). **Valve Gear**: Walschaerts. Piston valves.
Tractive Effort: 26290 lbf.

BR	*LMS*		
42968	13268–2968	Severn Valley Railway	Crewe 1934

CLASS 4MT 2-6-0

Built: 1947–52. Ivatt design. 162 built (43000–161).
Boiler Pressure: 225 lbf/sq in superheated. **Weight–Loco**: 59.1 tons.
Wheel Diameters: 3′ 0″, 5′ 3″. **–Tender**: 40.3 tons.
Cylinders: 17½″ x 26″ (O). **Valve Gear**: Walschaerts. Piston valves.
Tractive Effort: 24 170 lbf.

43106	Severn Valley Railway	Darlington 2148/1951

CLASS 4F 0-6-0

Built: 1911–41. Fowler Midland "Big Goods" design. Locomotives from 44027 onwards were LMS design with tenders with solid coal guards (Midland tenders had coal rails). The preserved Midland locomotive was given an LMS tender by BR. 772 built (43835–44606).
Boiler Pressure: 175 lbf/sq in superheated.
Wheel Diameter: 5′ 3″.
Cylinders: 20″ x 26″ (I).
Tractive Effort: 24560 lbf.
Weight–Loco: 48.75 tons.
–Tender: 41.2 tons
Valve Gear: Stephenson. Piston valves.

BR	*LMS*	*MR*		
43924	3924	3924	Keighley & Worth Valley Railway	Derby 1920
44027	4027		Gloucestershire Warwickshire Railway (N)	Derby 1924
44123	4123		Avon Valley Railway	Crewe 5658/1925
44422	4422		Nene Valley Railway	Derby 1927

CLASS 5MT "BLACK 5" 4-6-0

Built: 1934–51. Stanier design. 842 built (44658–45499).
Boiler Pressure: 225 lbf/sq in superheated.
Wheel Diameters: 3′ 3½″, 6′ 0″
Cylinders: 18½″ x 28″ (O).
Weight–Loco: 72.1 (* 75.3) tons.
–Tender: 53.65 (* 53.8) tons.
Tractive Effort: 25450 lbf.
Valve Gear: Walschaerts. Piston valves. 44767 has outside Stephenson with piston valves.

x Dual (air/vacuum) brakes.

BR	*LMS*			
44767*	4767	"GEORGE STEPHENSON"	North Yorkshire Moors Railway	Crewe 1947
44806	4806	"KENNETH ALDCROFT"	Llangollen Railway	Derby 1944
44871	4871		East Lancashire Railway	Crewe 1945
44901	4901		Barry Rail Centre	Crewe 1945
44932	4932		West Coast Railway Company, Carnforth	Horwich 1945
45000	5000		National Railway Museum, York (N)	Crewe 216/1935
45025	5025		Strathspey Railway	VF 4570/1934
45110	5110		Barrow Hill Roundhouse	VF 4653/1935
45163	5163		Colne Valley Railway	AW 1204/1935
45212	5212		North Yorkshire Moors Railway	AW 1253/1935
45231	5231	"THE SHERWOOD FORESTER"	East Lancashire Railway	AW 1286/1936
45293	5293		Colne Valley Railway	AW 1348/1936
45305	5305	"ALDERMAN AE DRAPER"	Great Central Railway	AW 1360/1937
45337	5337		East Lancashire Railway	AW 1392/1937
45379	5379		Mid Hants Railway	AW 1434/1937
45407x	5407	"THE LANCASHIRE FUSILIER"	East Lancashire Railway	AW 1462/1937
45428	5428	"ERIC TREACY"	North Yorkshire Moors Railway	AW 1483/1937
45491	5491		Great Central Railway	Derby 1943

CLASS 6P (Formerly 5XP) JUBILEE 4-6-0

Built: 1934–36. Stanier taper boiler development of Patriot class. 191 built (45552–45742).
Boiler Pressure: 225 lbf/sq in superheated.
Wheel Diameters: 3′ 3½″, 6′ 9″.
Cylinders: 17″ x 26″ (3).
Tractive Effort: 26610 lbf.
Weight–Loco: 79.55 tons.
–Tender: 53.65 tons.
Valve Gear: Walschaerts. Piston valves.

* Fitted with double chimney.

BR	*LMS*			
45593	5593	KOLHAPUR	Tyseley Locomotive Works	NBL 24151/1934
45596*	5596	BAHAMAS	Keighley & Worth Valley Railway	NBL 24154/1935
45690	5690	LEANDER	West Coast Railway Company, Carnforth	Crewe 288/1936
45699	5699	GALATEA	West Coast Railway Company, Carnforth	Crewe 297/1936

▲ 5MT 4-6-0 44767 "GEORGE STEPHENSON" shows off its unique Stephenson link motion at Kingsley & Froghall station on the Churnet Valley Railway on 19 November 2010. **Paul Abell**

▼ 6P Jubilee 4-6-0 5690 "LEANDER" climbs past Birkett Common with a returning "Fellsman" charter from Carlisle to Lancaster on 15 September 2010. **Hugh Ballantyne**

▲ Carrying BR Green livery, Rebuilt Royal Scot 4-6-0 46115 "SCOTS GUARDSMAN" is seen near Armathwaite with a Carlisle–York "Fellsman" railtour on 5 August 2009. **Brian Dobbs**

▼ Princess Royal 4-6-2 6201 "PRINCESS ELIZABETH" approaches Greenfield with the Railway Touring Company Crewe–Scarborough "Scarborough Flyer" on 27 August 2010. **Les Nixon**

CLASS 7P (Formerly 6P) ROYAL SCOT 4-6-0

Built: 1927–30. Fowler parallel design. All rebuilt 1943–55 with taper boilers and curved smoke deflectors. 71 built (46100–70).
Boiler Pressure: 250 lbf/sq in superheated.
Wheel Diameters: 3′ 3½″, 6′ 9″.
Cylinders: 18″ x 26″ (3).
Tractive Effort: 33 150 lbf.
Weight–Loco: 83 tons.
–Tender: 54.65 tons.
Valve Gear: Walschaerts. Piston valves.

BR	*LMS*			
46100	6100	ROYAL SCOT	Crewe Heritage Centre (N)	Derby 1930 reb Crewe 1950
46115	6115	SCOTS GUARDSMAN	West Coast Railway Company, Carnforth	NBL 23610/1927 reb Crewe 1947

6100 was built as 6152 THE KING'S DRAGOON GUARDSMAN. This loco swapped identities permanently with 6100 ROYAL SCOT in 1933 for a tour of the USA.

CLASS 8P (Formerly 7P) PRINCESS ROYAL 4-6-2

Built: 1933–35. Stanier design. 13 built (46200–12).
Boiler Pressure: 250 lbf/sq in superheated.
Wheel Diameters: 3′ 0″, 6′ 6″, 3′ 9″.
Cylinders: 16¼″ x 28″ (4).
Tractive Effort: 40 290 lbf.
Weight–Loco: 105.5 tons.
–Tender: 54.65 tons.
Valve Gear: Walschaerts. Piston valves.

x Dual (air/vacuum) brakes.

BR	*LMS*			
46201 x	6201	PRINCESS ELIZABETH	Crewe Heritage Centre	Crewe 107/1933
46203	6203	PRINCESS MARGARET ROSE	Midland Railway-Butterley	Crewe 253/1935

CLASS 8P (Formerly 7P) PRINCESS CORONATION 4-6-2

Built: 1937–48. Stanier design. 24 of this class were built streamlined but had the casing removed later. Certain locomotives were built with single chimneys, but all finished up with double chimneys. The tenders were fitted with steam driven coal-pushers. 38 built (46220–57).
Boiler Pressure: 250 lbf/sq in superheated.
Wheel Diameters: 3′ 0″, 6′ 9″, 3′ 9″.
Cylinders: 16½″ x 28″ (4).
Tractive Effort: 40 000 lbf.
Weight–Loco: 105.25 tons.
–Tender: 56.35 tons.
Valve Gear: Walschaerts. Piston valves.

[1] Built streamlined and with single chimney. Double chimney fitted 1943. Destreamlined 1947–2009.
[2] Never streamlined and built with single chimney. Double chimney fitted 1941. Dual (air/vacuum) brakes.
[3] Built streamlined and with double chimney. Destreamlined 1946.

BR	*LMS*			
46229[1]	6229	DUCHESS OF HAMILTON	National Railway Museum, York (N)	Crewe 1938
46233[2]	6233	DUCHESS OF SUTHERLAND	Midland Railway-Butterley	Crewe 1938
46235[3]	6235	CITY OF BIRMINGHAM	Thinktank: Birmingham Science Museum	Crewe 1939

6229 was numbered 6220 whilst in the USA between 1938 and 1943.

CLASS 2MT — 2-6-0

Built: 1946–53. Ivatt design. 128 built (46400–527).
Boiler Pressure: 200 lbf/sq in superheated.
Wheel Diameters: 3′ 0″, 5′ 0″.
Cylinders: 16″ (* 16½″) x 24″ (O).
Tractive Effort: 17410 (* 18510) lbf.
Weight–Loco: 47.1 (* 48.45) tons.
–Tender: 37.15 tons.
Valve Gear: Walschaerts. Piston valves.

46428		East Lancashire Railway	Crewe 1948
46441		Ribble Steam Railway	Crewe 1950
46443		Severn Valley Railway	Crewe 1950
46447		Isle of Wight Steam Railway	Crewe 1950
46464		Caledonian Railway	Crewe 1950
46512*	"E.V. COOPER ENGINEER"	Strathspey Railway	Swindon 1952
46521*		Great Central Railway	Swindon 1953

CLASS 3F — "JINTY" — 0-6-0T

Built: 1924–31. Fowler LMS development of Midland design. 422 built (47260–47681).
Boiler Pressure: 160 lbf/sq in.
Wheel Diameter: 4′ 7″.
Valve Gear: Stephenson. Slide valves.
Weight: 49.5 tons.
Cylinders: 18″ x 26″ (I).
Tractive Effort: 20830 lbf.

BR	*LMS*		
47279	7119–7279	Keighley & Worth Valley Railway	VF 3736/1924
47298	7138–7298	Llangollen Railway	HE 1463/1924
47324	16407–7324	Midland Railway-Butterley	NBL 23403/1926
47327	16410–7327	Midland Railway-Butterley	NBL 23406/1926
47357	16440–7357	Midland Railway-Butterley	NBL 23436/1926
47383	16466–7383	Severn Valley Railway	VF 3954/1926
47406	16489–7406	Great Central Railway	VF 3977/1926
47445	16528–7445	Midland Railway-Butterley	HE 1529/1927
47493	16576–7493	Spa Valley Railway	VF 4195/1928
47564	16647–7564	Midland Railway-Butterley	HE 1580/1928

47324 is on loan from the East Lancashire Railway.
47564 was latterly used as a stationary boiler numbered 2022.

CLASS 8F — 2-8-0

Built: 1934–46. Stanier design. 331 built for LMS (8000–8225, 8301–8399, 8490–8495). A further 521 were built to Ministry of Supply (208) Railway Executive Committee (245) and LNER (68) orders. Many of these operated on Britain's Railways with 228 being shipped overseas during the war. Post-war many were taken into LMS/BR stock including some returned from overseas.
Boiler Pressure: 225 lbf/sq in superheated.
Wheel Diameters: 3′ 3½″, ,4′ 8½″.
Cylinders: 18½″ x 28″ (O).
Tractive Effort: 32440 lbf.
Weight–Loco: 72.1 tons.
–Tender: 53.65 tons.
Valve Gear: Walschaerts. Piston valves.

* Number allocated but never carried.
§ Became Persian Railways 41.109.

BR	*LMS*	*WD*		
48151	8151		West Coast Railway Company, Carnforth	Crewe 1942
48173	8173		Churnet Valley Railway	Crewe 1943
48305	8305		Great Central Railway	Crewe 1943
48431	8431		Keighley & Worth Valley Railway	Swindon 1944
48624	8624		Great Central Railway	Ashford 1943
48773	8233	307–70307–500§	Severn Valley Railway	NBL 24607/1940

48151 has been named "GAUGE O' GUILD".
48518 has been dismantled at the Llangollen Railway and the boiler used for the replica 1014 COUNTY OF GLAMORGAN.

In addition the following three locos were built to Ministry of Supply orders and have been repatriated to Great Britain:

Dual (air/vacuum) brakes.

BR	*LMS*	*WD*	*TCDD*		
–	8274*	348	45160	Gloucestershire Warwickshire Railway	NBL 24648/1940
–	8267*	341	45166	Barry Rail Centre	NBL 24641/1940
–		554	45170	National Railway Museum, Shildon	NBL 24755/1942

LNWR CLASS G2 (7F) — 0-8-0

Built: 1921–22. Beames development of earlier Bowen-Cooke LNWR design. 60 built (49395–454). In addition many earlier locos were rebuilt to similar condition.
Boiler Pressure: 175 lbf/sq in superheated.
Wheel Diameter: 4′ 5½″.
Cylinders: 20½″ x 24″ (I).
Tractive Effort: 28040 lbf.
Weight–Loco: 62 tons.
–Tender: 40.75 tons.
Valve Gear: Joy. Piston valves.

BR	*LMS*	*LNWR*		
49395	9395	485	North Yorkshire Moors Railway (N)	Crewe 5662/1921

L&Y CLASS 5 (2P) — 2-4-2T

Built: 1889–1909. Aspinall L&Y design. 309 built.
Boiler Pressure: 180 lbf/sq in.
Wheel Diameters: 3′ 7⅞″, 5′ 8″, 3′ 7⅞″.
Valve Gear: Joy. Slide valves.
Weight: 55.45 tons.
Cylinders: 18″ x 26″ (I).
Tractive Effort: 18990 lbf.

BR	*LMS*	*L&Y*		
50621	10621	1008	National Railway Museum, York (N)	Horwich 1/1889

L&Y CLASS 21 (0F) "PUG" — 0-4-0ST

Built: 1891–1910. Aspinall L&Y design. 57 built.
Boiler Pressure: 160 lbf/sq in.
Wheel Diameter: 3′ 0¾″.
Valve Gear: Stephenson. Slide valves.
Weight: 21.25 tons.
Cylinders: 13″ x 18″ (O).
Tractive Effort: 11370 lbf.

BR	*LMS*	*L&Y*		
51218	11218	68	Keighley & Worth Valley Railway	Horwich 811/1901
	11243	19	Ribble Steam Railway	Horwich 1097/1910

L&Y CLASS 23 (2F) — 0-6-0ST

Built: 1891–1900. Aspinall rebuild of Barton Wright L&Y 0-6-0. 230 rebuilt.
Boiler Pressure: 140 lbf/sq in.
Wheel Diameter: 4′ 6″.
Valve Gear: Joy. Slide valves.
Weight: 43.85 tons
Cylinders: 17½″ x 26″ (I).
Tractive Effort: 17590 lbf.

BR	*LMS*	*L&Y*		
–	11456	752	Keighley & Worth Valley Railway	BP 1989/1881 reb. Hor. 1896

L&Y CLASS 25 (2F) — 0-6-0

Built: 1876–87. Barton Wright L&Y design. 280 built.
Boiler Pressure: 140 lbf/sq in.
Wheel Diameter: 4′ 6″.
Cylinders: 17½″ x 26″ (I).
Tractive Effort: 17590 lbf.
Weight–Loco: 39.05 tons.
–Tender: 28.5 tons.
Valve Gear: Joy. Slide valves.

BR	*LMS*	*L&Y*		
52044	12044	957	Keighley & Worth Valley Railway	BP 2840/1887

L&Y CLASS 27 (3F) — 0-6-0

Built: 1889–1917. Aspinall L&Y design. 448 built.
Boiler Pressure: 180 lbf/sq in.
Wheel Diameter: 5′ 1″.
Cylinders: 18″ x 26″ (I).
Tractive Effort: 21 170 lbf.
Weight–Loco: 44.3 tons.
–Tender: 26.1 tons.
Valve Gear: Joy. Slide valves.

BR	*LMS*	*L&Y*		
52322	12322	1300	Ribble Steam Railway	Horwich 420/1896

CLASS 7F — 2-8-0

Built: 1914–25. Fowler design for Somerset & Dorset Joint Railway (Midland and LSWR jointly owned). 11 built (53800–10).
Boiler Pressure: 190 lbf/sq in superheated.
Wheel Diameters: 3′ 3½″, 4′ 7½″.
Cylinders: 21″ x 28″ (O).
Tractive Effort: 35 950 lbf.
Weight–Loco: 64.75 tons.
–Tender: 26.1 tons.
Valve Gear: Walschaerts. Piston valves.

BR	*LMS*	*S&DJR*		
53808	9678–13808	88	West Somerset Railway	RS 3894/1925
53809	9679–13809	89	North Yorkshire Moors Railway	RS 3895/1925

CALEDONIAN RAILWAY (1P) — 4-2-2

Built: 1886. Drummond design. 1 built for the Edinburgh International Exhibition.
Boiler Pressure: 160 lbf/sq in.
Wheel Diameters: 3′ 6″, 7′ 0″, 4′ 6″.
Cylinders: 18″ x 26″ (I).
Tractive Effort: 13 640 lbf.
Weight–Loco: 41.35 tons.
–Tender: 35.4 tons.
Valve Gear: Stephenson. Slide valves.

BR	*LMS*	*CR*		
–	14010	123	Glasgow Riverside Museum	N 3553/1886

CALEDONIAN RAILWAY 439 CLASS (2P) — 0-4-4T

Built: 1900–14. McIntosh design. 68 built.
Boiler Pressure: 160 lbf/sq in.
Wheel Diameters: 5′ 9″, 3′ 2″.
Valve Gear: Stephenson. Slide valves.
Weight: 53.95 tons.
Cylinders: 18″ x 26″ (I).
Tractive Effort: 16 600 lbf.

Dual (air/vacuum) brakes.

BR	*LMS*	*CR*		
55189	15189	419	Bo'ness & Kinneil Railway	St Rollox 1907

GSWR 322 CLASS (3F) — 0-6-0T

Built: 1917. Drummond design. 3 built.
Boiler Pressure: 160 lbf/sq in.
Wheel Diameter: 4′ 2″.
Valve Gear: Walschaerts. Piston valves.
Weight: 40 tons.
Cylinders: 17″ x 22″ (O).
Tractive Effort: 17 290 lbf.

BR	*LMS*	*GSWR*		
–	16379	9	Glasgow Riverside Museum	NBL 21521/1917

CALEDONIAN RAILWAY 812 CLASS (3F) — 0-6-0

Built: 1899–1900. McIntosh design. 96 built (57550–645).
Boiler Pressure: 160 lbf/sq in.
Wheel Diameter: 5′ 0″.
Cylinders: 18½″ x 26″ (I).
Tractive Effort: 20 170 lbf.
Weight–Loco: 45.7 tons.
–Tender: 37.9 tons.
Valve Gear: Stephenson. Slide valves.

Air brakes.

BR	*LMS*	*CR*		
57566	17566	828	Strathspey Railway	St. Rollox 1899

▲ Visiting from the West Somerset Railway, Somerset & Dorset 7F 2-8-0 No. 88 calls at Medstead & Four Marks with the 10.05 Alresford–Alton on 18 September 2010. **Phil Barnes**

▼ 3F 0-6-0T 47279 climbs the Worth Valley near Oakworth with a train from Keighley to Oxenhope on 15 November 2009. **Brian Dobbs**

HIGHLAND RAILWAY (4F) "JONES GOODS" 4-6-0

Built: 1894. Jones design. 15 built.
Boiler Pressure: 175 lbf/sq in.
Wheel Diameters: 3′ 3″, 5′ 3″.
Cylinders: 20″ x 26″ (O).
Tractive Effort: 24560 lbf.
Weight–Loco: 56 tons.
–Tender: 38.35 tons.
Valve Gear: Stephenson. Slide valves.

BR	*LMS*	*HR*		
–	17916	103	Glasgow Riverside Museum	SS 4022/1894

NORTH LONDON RAILWAY 75 CLASS (2F) 0-6-0T

Built: 1879–1905. Park design. 30 built.
Boiler Pressure: 160 lbf/sq in.
Wheel Diameter: 4′ 4″.
Valve Gear: Stephenson. Slide valves.
Weight: 45.55 tons.
Cylinders: 17″ x 24″ (O).
Tractive Effort: 18140 lbf.

BR	*LMS*	*LNWR*	*NLR*		
58850	7505–27505	2650	76–116	Bluebell Railway	Bow 181/1881

LNWR COAL TANK (2F) 0-6-2T

Built: 1881–1897. Webb design. 300 built.
Boiler Pressure: 150 lbf/sq in.
Wheel Diameters: 4′ 5½″, 3′ 9″.
Valve Gear: Stephenson. Slide valves.
Weight: 43.75 tons.
Cylinders: 17″ x 24″ (I).
Tractive Effort: 16530 lbf.

BR	*LMS*	*LNWR*		
58926	7799	1054	Keighley & Worth Valley Railway	Crewe 2979/1888

MIDLAND 156 CLASS (1P) 2-4-0

Built: 1866–68. Kirtley design. 31 built.
Boiler Pressure: 140 lbf/sq in.
Wheel Diameters: 4′ 3″, 6′ 3″.
Cylinders: 18″ x 24″ (I).
Tractive Effort: 12340 lbf.
Weight–Loco: 41.25 tons.
–Tender: 34.85 tons.
Valve Gear: Stephenson. Slide valves.

BR	*LMS*	*MR*		
–	2–20002	158–158A–2	Midland Railway-Butterley (N)	Derby 1866

NORTH STAFFS RAILWAY New L CLASS (3F) 0-6-2T

Built: 1903–23. Hookham design. 34 built.
Boiler Pressure: 175 lbf/sq in.
Wheel Diameters: 5′ 0″, 4′ 0″.
Valve Gear: Stephenson. Slide valves.
Weight: 64.95 tons.
Cylinders: 18½″ x 26″ (I).
Tractive Effort: 22060 lbf.

LMS	*NSR*		
2271	2	National Railway Museum, Shildon (N)	Stoke 1923

Although built in 1923, this loco carried an NSR number, since the NSR was not taken over by the LMS until 1 July 1923.

LNWR PRECEDENT (1P) 2-4-0

Built: 1874–82 (renewed 1887–1901). Webb design. 166 built.
Boiler Pressure: 150 lbf/sq in.
Wheel Diameters: 3′ 9″, 6′ 9″.
Cylinders: 17″ x 24″ (I).
Tractive Effort: 10920 lbf.
Weight–Loco: 35.6 tons.
–Tender: 25 tons.
Valve Gear: Allan.

LMS	*LNWR*			
5031	790	HARDWICKE	National Railway Museum, York (N)	Crewe 3286/1892

LNWR 2-2-2

Built: 1847. Trevithick design rebuilt by Ramsbottom in 1858.
Boiler Pressure: 140 lbf/sq in.
Wheel Diameters: 3′ 6″, 8′ 6″, 3′ 6″.
Cylinders: 17¼″ x 24″ (O).
Tractive Effort: 8330 lbf
Weight–Loco: 29.9 tons.
–Tender: 25 tons.
Valve Gear: Stephenson. Slide valves.

LNWR			
173–3020	CORNWALL	National Railway Museum, Shildon (N)	Crewe 35/1847

LNWR 0-4-0ST

Built: 1865. Ramsbottom design.
Boiler Pressure: 120 lbf/sq in.
Wheel Diameter: 4′ 0″.
Tractive Effort: 8330 lbf.
Weight: 22.75 tons.
Cylinders: 14″ x 20″ (I).

LNWR		
1439–1985–3042	Ribble Steam Railway (N)	Crewe 842/1865

GRAND JUNCTION RAILWAY 2-2-2

Built: 1845. Trevithick design.
Boiler Pressure: 120 lbf/sq in.
Wheel Diameters: 3′ 6″, 6′ 0″, 3′ 6″.
Cylinders: 15″ x 20″ (O).
Tractive Effort: 6375 lbf.
Weight–Loco: 20.4 tons.
–Tender: 16.4 tons.
Valve Gear: Allan.

LNWR	*GJR*			
49	49–1868	COLUMBINE	Science Museum, London (N)	Crewe 25/1845

FURNESS RAILWAY 0-4-0

Built: 1846.
Boiler Pressure: 110 lbf/sq in.
Wheel Diameter: 4′ 9″.
Cylinders: 14″ x 24″ (I).
Tractive Effort: 7720 lbf.
Weight–Loco: 20 tons.
–Tender: 13 tons.
Valve Gear: Stephenson. Slide valves.

3 COPPERNOB National Railway Museum, York (N) BCK 1846

FURNESS RAILWAY 0-4-0

Built: 1863 as 0-4-0. Sold in 1870 to Barrow Steelworks and numbered 7. Rebuilt 1915 as 0-4-0ST. Restored to original condition 1999.
Boiler Pressure: 120 lbf/sq in.
Cylinders: 15½″ x 24″(I).
Wheel Diameter: 4′10″.
Tractive Effort: 10 140 lbf.

20 National Railway Museum, Shildon SS 1435/1863

FURNESS RAILWAY 0-4-0ST

Built: 1865 as 0-4-0. Sold in 1873 to Barrow Steelworks and numbered 17. Rebuilt 1921 as 0-4-0ST.
Boiler Pressure: 120 lbf/sq in.
Cylinders: 15½″ x 24″(I).
Wheel Diameter: 4′3″.
Tractive Effort: 11 532 lbf.

FR	*Present*		
25	6	West Coast Railway Company, Carnforth	SS 1585/1865

▲ One of the 8F 2-8-0s brought back from Turkey, 8274 (45160) passes Little Woodthorpe with the 10.40 Loughborough–Rothley demonstration freight on 10 April 2011. **Paul Biggs**

▼ Although none of the 8Fs was ever painted LMS Red, 8624 has been turned out in this livery as a "what might have been". On 24 May 2009 it approaches Darley Dale with a train from Matlock Riverside to Rowsley. This loco is now based at the Great Central Railway. **Les Nixon**

LIVERPOOL & MANCHESTER RAILWAY 0-2-2

Built: 1829 for the Rainhill trials.
Boiler Pressure: 50 lbf/sq in.
Wheel Diameters: 4′ 8½″, 2′ 6″.
Cylinders: 8″ x 17″ (O).
Weight–Loco: 4.25 tons.
–Tender: 5.2 tons.
Tractive Effort: 820 lbf.

ROCKET	Science Museum, London (N)	RS 1/1829

LIVERPOOL & MANCHESTER RAILWAY 0-4-0

Built: 1829 for the Rainhill trials.
Boiler Pressure: 50 lbf/sq in.
Wheel Diameter: 4′ 6″.
Cylinders: 7″ x 18″ (O).
Weight–Loco: 4.25 tons.
–Tender: 5.2 tons.
Tractive Effort: 690 lbf.

SANS PAREIL	National Railway Museum, Shildon (N)	Hack 1829

LIVERPOOL & MANCHESTER RAILWAY 0-4-2

Built: 1838–39. 4 built. The survivor was the star of the film "The Titfield Thunderbolt".
Boiler Pressure: 50 lbf/sq in.
Wheel Diameters: 5′ 0″, 3′ 3″.
Cylinders: 14″ x 24″ (I).
Weight–Loco: 14.45 tons.
–Tender:
Tractive Effort: 3330 lbf.

L&MR	*LNWR*			
57	116	LION	Liverpool Museum Store, Bootle	TKL 1838

MERSEY RAILWAY 0-6-4T

Built: 1885. Withdrawn 1903 on electrification. No. 5 was sold to Shipley colliery in Derbyshire.
Boiler Pressure: 150 lbf/sq in.
Wheel Diameters: 4′ 7″, 3′ 0″.
Valve Gear: Stephenson. Slide valves.
Weight: 67.85 tons.
Cylinders: 21″ x 26″ (I).
Tractive Effort: 26 600 lbf.

1	THE MAJOR	Rail Transport Museum, Thirlmere, NSW, Australia	BP 2601/1885
5	CECIL RAIKES	Liverpool Museum Store, Bootle	BP 2605/1885

HIGHLAND RAILWAY (DUKE OF SUTHERLAND) 0-4-4T

Built: 1895 for the Duke of Sutherland. Exported to Canada 1965, repatriated to Britain 2011.
Boiler Pressure: 150 lbf/sq in.
Wheel Diameters: 4′ 6″, 3′ 0″.
Valve Gear: Stephenson.
Weight: 25 tons.
Cylinders: 21″ x 26″ (I).
Tractive Effort: 7183 lbf.

–	DUNROBIN	North of England Open Air Museum, Beamish	SS 4085/1895

▶ Spring 2011 saw the shipment back to Britain from Canada of the Duke of Sutherland's personal 0-4-4T "DUNROBIN" (used by him on the Highland Railway, for which his family had provided much of the finance). The loco has been bought by the North of England Open Air Museum at Beamish, but it was first moved to the Severn Valley Railway workshops at Bridgnorth for an assessment of the work required to return it to working order. On 20 May it is being carefully unloaded at Bridgnorth after its long journey. **Paul Jarman**

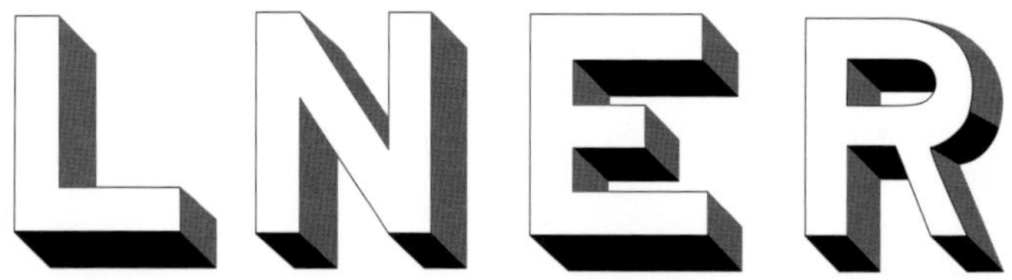

1.4. LONDON & NORTH EASTERN RAILWAY AND CONSTITUENT COMPANIES' STEAM LOCOMOTIVES

GENERAL

The LNER was formed in 1923 by the amalgamation of the Great Northern Railway (GNR). North Eastern Railway (NER), Great Eastern Railway (GER), Great Central Railway (GCR), North British Railway (NBR) and Great North of Scotland Railway (GNSR). Prior to this the Hull & Barnsley Railway (H&B) had been absorbed by the NER in 1922.

The newly-built Class A1 Pacific 60163 TORNADO can be found in the "New Construction" section on page 74.

LOCOMOTIVE NUMBERING SYSTEM

Initially pre grouping locomotive numbers were retained, but in September 1923 suffix letters started to be applied depending upon the works which repaired the locomotives. In 1924 locomotives were renumbered in blocks as follows: NER locomotives remained unaltered, GNR locomotives had 3000 added, GCR 5000, GNSR 6800, GER 7000 and NBR 9000. New locomotives filled in gaps between existing numbers. By 1943 the numbering of new locomotives had become so haphazard that it was decided to completely renumber locomotives so that locomotives of a particular class were all contained in the same block of numbers. This was carried out in 1946. On nationalisation in 1948, 60000 was added to LNER numbers.

CLASSIFICATION SYSTEM

The LNER gave each class a unique code consisting of a letter denoting the wheel arrangement and a number denoting the individual class within the wheel arrangement. Route availability (RA) was denoted by a number, the higher the number the more restricted the route availability.

▲ In early 2011 A4 4-6-2 60019 "BITTERN" emerged in LNER Garter Blue as 4492 "DOMINION OF NEW ZEALAND", with the pre-War valances restored over the driving wheels. On an early trip in its new guise, 4492 passes through Heeley in Sheffield with a York–Stratford-upon-Avon charter on 12 May 2011. Note the second tender used to provide extra water capacity. **Steve Smith**

▼ Newly outshopped in wartime black, A3 4-6-2 103 "FLYING SCOTSMAN" takes pride of place on the turntable in the Great Hall at NRM York on 27 May 2011. **Fred Kerr**

CLASS A4 4-6-2

Built: 1935–38. Gresley streamlined design. "MALLARD" attained the world speed record for a steam locomotive of 126mph in 1938 and is still unbeaten. 35 built (2509–12, 4462–69/82–4500/4900–03).
Boiler Pressure: 250 lbf/sq in superheated. **Weight–Loco:** 102.95 tons.
Wheel Diameters: 3′ 2″, 6′ 8″, 3′ 8″. **–Tender:** 64.15 tons.
Cylinders: 18½″ x 26″ (3). **Tractive Effort:** 35450 lbf.
Valve Gear: Walschaerts with derived motion for inside cylinder. Piston valves.
BR Power Classification: 8P. **RA:** 9.

x Dual (air/vacuum) brakes.

BR	*LNER*			
60007x	4498–7	SIR NIGEL GRESLEY	North Yorkshire Moors Railway	Doncaster 1863/1937
60008	4496–8	DWIGHT D. EISENHOWER	National RR Museum, USA	Doncaster 1861/1937
60009x	4488–9	UNION OF SOUTH AFRICA	Thornton Depot, Fife	Doncaster 1853/1937
60010	4489–10	DOMINION OF CANADA	Canadian RR Historical Museum	Doncaster 1854/1937
60019	4464–19	BITTERN	Southall Depot, London	Doncaster 1866/1937
60022	4468–22	MALLARD	National Railway Museum, Shildon (N)	Doncaster 1870/1938

60019 sometimes operates with a second tender for carrying water. Currently running as 4492 DOMINION OF NEW ZEALAND.

CLASS A3 4-6-2

Built: 1922–35. Gresley design. Built as Class A1 (later reclassified A10 after the Peppercorn A1s were being designed), but rebuilt to A3 in 1947. 79 built. (60035–113). Rebuilt (1959) with Kylchap blastpipe, double chimney and (1961) German-style smoke deflectors. Restored to single chimney on preservation in 1963, but restored to rebuilt state in 1993. The loco is numbered 4472, a number it never carried as an A3, since it was renumbered 103 in 1946.
Boiler Pressure: 220 lbf/sq in superheated. **Weight–Loco:** 96.25 tons.
Wheel Diameters: 3′ 2″, 6′ 8″, 3′ 8″. **–Tender:** 62.4 tons.
Cylinders: 19″ x 26″ (3). **Tractive Effort:** 32910 lbf.
Valve Gear: Walschaerts with derived motion for inside cylinder. Piston valves.
BR Power Classification: 7P. **RA:** 9.

Dual (air/vacuum) brakes.

BR	*LNER*			
60103	1472–4472–502–103	FLYING SCOTSMAN	National Railway Museum, York (N)	Doncaster 1564/1923

CLASS A2 4-6-2

Built: 1947–48. Peppercorn development of Thompson design. 15 built.
Boiler Pressure: 250 lbf/sq in superheated. **Weight–Loco:** 101 tons.
Wheel Diameters: 3′ 2″, 6′ 2″, 3′ 8″. **–Tender:** 60.35 tons.
Cylinders: 19″ x 26″ (3). **Valve Gear:** Walschaerts. Piston valves.
Tractive Effort: 40430 lbf. **BR Power Classification:** 8P.
RA: 9.

60532	BLUE PETER	Barrow Hill Roundhouse	Doncaster 2023/1948

CLASS V2 2-6-2

Built: 1936–44. Gresley design for express passenger and freight. 184 built (60800–983).
Boiler Pressure: 220 lbf/sq in superheated. **Weight–Loco:** 93.1 tons.
Wheel Diameters: 3′ 2″, 6′ 2″, 3′ 8″. **–Tender:** 52 tons.
Cylinders: 18½″ x 26″ (3). **Tractive Effort:** 33730 lbf.
Valve Gear: Walschaerts with derived motion for inside cylinder. Piston valves.
BR Power Classification: 6MT. **RA:** 9.

BR	*LNER*			
60800	4771–800	GREEN ARROW	National Railway Museum, York (N)	Doncaster 1837/1936

CLASS B1 4-6-0

Built: 1942–52. Thompson design. 410 built (61000–409).
Boiler Pressure: 225 lbf/sq in superheated.
Wheel Diameters: 3′ 2″, 6′ 2″.
Cylinders: 20″ x 26″ (O).
Tractive Effort: 26880 lbf.
RA: 5.
Weight–Loco: 71.15 tons.
–Tender: 52 tons.
Valve Gear: Walschaerts. Piston valves.
BR Power Classification: 5MT.

BR	*LNER*	*Present*		
61264	1264		Barrow Hill Roundhouse	NBL 26165/1947
61306		1306 "MAYFLOWER"	Battlefield Railway	NBL 26207/1948

The original MAYFLOWER was 61379. 61264 also carried Departmental 29.

CLASS B12 4-6-0

Built: 1911–28. Holden GER design. 81 built (1500–70, 8571–80). GER Class S69.
61572 Rebuilt to B12/3 1933.
Boiler Pressure: 180 lbf/sq in superheated.
Wheel Diameters: 3′ 3″, 6′ 6″.
Cylinders: 20″ x 28″ (I).
Tractive Effort: 21970 lbf.
RA: 5.
Weight –Loco: 69.5 tons.
–Tender: 39.3 tons.
Valve Gear: Stephenson. Piston valves.
BR Power Classification: 4P.

Dual (air/vacuum) brakes.

BR	*LNER*			
61572	8572	1572	North Norfolk Railway	BP 6488/1928

CLASS K4 2-6-0

Built: 1937–38. Gresley design for West Highland line. 6 built (61993–8).
Boiler Pressure: 200 lbf/sq in superheated.
Wheel Diameters: 3′ 2″, 5′ 2″.
Cylinders: 18½″ x 26″ (3).
Weight–Loco: 68.4 tons.
–Tender: 44.2 tons.
Tractive Effort: 36 600 lbf.
Valve Gear: Walschaerts with derived motion for inside cylinder. Piston valves.
BR Power Classification: 5P6F.
RA: 6.

BR	*LNER*			
61994	3442–1994	THE GREAT MARQUESS	Thornton Depot, Fife	Darlington 1761/1938

CLASS K1 2-6-0

Built: 1949–50. Peppercorn design. 70 built (62001–70).
Boiler Pressure: 225 lbf/sq in superheated.
Wheel Diameters: 3′ 2″, 5′ 2″.
Cylinders: 20″ x 26″ (O).
Tractive Effort: 32080 lbf.
RA: 6.
Weight–Loco: 66 tons.
–Tender: 52.2 tons.
Valve Gear: Walschaerts. Piston valves.
BR Power Classification: 6MT.

BR	*Present*		
62005	2005	North Yorkshire Moors Railway	NBL 26609/1949

CLASS D40 4-4-0

Built: 1899–1921. Pickersgill GNSR Class F. 21 built.
Boiler Pressure: 165 lbf/sq in superheated.
Wheel Diameters: 3′ 9½″, 6′ 1″.
Cylinders: 18″ x 26″ (I).
Tractive Effort: 16180 lbf.
RA: 4.
Weight –Loco: 48.65 tons.
–Tender: 37.4 tons.
Valve Gear: Stephenson. Slide valves.
BR Power Classification: 1P.

BR	*LNER*	*GNSR*			
62277	6849–2277	49	GORDON HIGHLANDER	Bo'ness & Kinneil Railway	NBL 22563/1920

▲ K4 2-6-0 61994 "THE GREAT MARQUESS" is near Loch Gowan climbing from Achnasheen to Luib with the Inverness–Kyle of Lochalsh leg of the "Great Britain III" tour on 12 April 2010.
Jon Bowers

▼ Making a rare visit to England, J36 0-6-0 65243 "MAUDE" stands on the turntable at NRM York on 12 April 2011, with No. 66 "AEROLITE" visible to the left. **Paul Abell**

CLASS D34 GLEN 4-4-0

Built: 1913–20. Reid NBR Class K. 32 built.
Boiler Pressure: 165 lbf/sq in superheated.
Wheel Diameters: 3′ 6″, 6′ 0″.
Cylinders: 20″ x 26″ (I).
Tractive Effort: 22 100 lbf.
RA: 6.
Weight–Loco: 57.2 tons.
–Tender: 46.65 tons.
Valve Gear: Stephenson. Piston valves.
BR Power Classification: 3P.

BR	*LNER*	*NBR*			
62469	9256–2469	256	GLEN DOUGLAS	Glasgow Riverside Museum	Cowlairs 1913

CLASS D11 IMPROVED DIRECTOR 4-4-0

Built: 1919–22. Robinson GCR Class 11F. 11 built (62660–70). 24 similar locos were built by the LNER.
Boiler Pressure: 180 lbf/sq in superheated.
Wheel Diameters: 3′ 6″, 6′ 9″.
Cylinders: 20″ x 26″ (I).
Tractive Effort: 19 640 lbf.
RA: 6.
Weight–Loco: 61.15 tons.
–Tender: 48.3 tons.
Valve Gear: Stephenson. Piston valves.
BR Power Classification: 3P.

BR	*LNER*	*GCR*			
62660	5506–2660	506	BUTLER HENDERSON	Barrow Hill Roundhouse (N)	Gorton 1919

CLASS D49/1 4-4-0

Built: 1927–29. Gresley design. 36 built (62700–35).
Boiler Pressure: 180 lbf/sq in superheated.
Wheel Diameters: 3′ 1¼″, 6′ 8″.
Cylinders: 17″ x 26″ (3).
Valve Gear: Walschaerts with derived motion for inside cylinder. Piston valves.
BR Power Classification: 4P.
Weight–Loco: 66 tons.
–Tender: 52 tons.
Tractive Effort: 21 560 lbf.
RA: 8.

BR	*LNER*			
62712	246–2712	MORAYSHIRE	Bo'ness & Kinneil Railway	Darlington 1391/1928

CLASS E4 2-4-0

Built: 1891–1902. Holden GER Class T26. 100 built.
Boiler Pressure: 160 lbf/sq in.
Wheel Diameters: 4′ 0″, 5′ 8″.
Cylinders: 17½″ x 24″ (I).
Tractive Effort: 14 700 lbf.
RA: 2.
Weight–Loco: 40.3 tons.
–Tender: 30.65 tons.
Valve Gear: Stephenson. Slide valves.
BR Power Classification: 1MT.

Air brakes.

BR	*LNER*	*GER*		
62785	7490–7802–2785	490	Bressingham Steam Museum (N)	Stratford 836/1894

CLASS C1 4-4-2

Built: 1902–10. Ivatt GNR Class C1. 94 built.
Boiler Pressure: 170 lbf/sq in.
Wheel Diameters: 3′ 8″, 6′ 8″, 3′ 8″.
Cylinders: 20″ x 24″ (O).
Tractive Effort: 17 340 lbf.
RA: 7.
Weight–Loco: 69.6 tons.
–Tender: 43.1 tons.
Valve Gear: Stephenson. Slide valves.
BR Power Classification: 2P.

BR	*LNER*	*GNR*		
–	3251–2800	251	Bressingham Steam Museum (N)	Doncaster 991/1902

CLASS Q6 0-8-0

Built: 1913–21. Raven NER Class T2. 120 built (63340–459).
Boiler Pressure: 180 lbf/sq in superheated.
Wheel Diameter: 4′ 7¼″.
Cylinders: 20″ x 26″ (O).
Tractive Effort: 28800 lbf.
RA: 6.
Weight–Loco: 65.9 tons.
–Tender: 44.1 tons.
Valve Gear: Stephenson. Piston valves.
BR Power Classification: 6F.

BR	*LNER*	*NER*		
63395	2238–3395	2238	North Yorkshire Moors Railway	Darlington 1918

CLASS Q7 0-8-0

Built: 1919–24. Raven NER Class T3. 15 built (63460–74).
Boiler Pressure: 180 lbf/sq in superheated.
Wheel Diameter: 4′ 7¼″.
Cylinders: 18½″ x 26″ (3).
Tractive Effort: 36960 lbf.
RA: 7.
Weight–Loco: 71.6 tons.
–Tender: 44.1 tons.
Valve Gear: Stephenson. Piston valves.
BR Power Classification: 8F.

BR	*LNER*	*NER*		
63460	901–3460	901	Head of Steam, Darlington Railway Museum (N)	Darlington 1919

CLASS O4 2-8-0

Built: 1911–20. Robinson GCR Class 8K. 129 built. A further 521 were built, being ordered by the Railway Operating Department (ROD). These saw service on British and overseas railways during and after World War I. Some subsequently passed to British railway administrations whilst others were sold abroad.
Boiler Pressure: 180 lbf/sq in superheated.
Wheel Diameters: 3′ 6″, 4′ 8″.
Cylinders: 21″ x 26″ (O).
Tractive Effort: 31330 lbf.
RA: 6.
Weight–Loco: 73.2 tons.
–Tender: 48.3 tons.
Valve Gear: Stephenson. Piston valves.
BR Power Classification: 7F.

BR	*LNER*	*GCR*		
63601	5102–3509–3601	102	Great Central Railway (N)	Gorton 1911

ROD		
1984	Dorrigo, Northern New South Wales, Australia	NBL 22042/1918
2003	Dorrigo, Northern New South Wales, Australia	Gorton 1918
2004	Richmond Vale Steam Centre, Kurri-Kurri, NSW, Australia	Gorton 1918

CLASS J21 0-6-0

Built: 1886–95. Worsdell NER Class C. 201 built.
Boiler Pressure: 160 lbf/sq in superheated.
Wheel Diameter: 5′ 1¼″.
Cylinders: 19″ x 24″ (I).
Tractive Effort: 19240 lbf.
RA: 3.
Weight–Loco: 43.75 tons.
–Tender: 36.95 tons.
Valve Gear: Stephenson. Piston valves.
BR Power Classification: 2F.

BR	*LNER*	*NER*		
65033	876–5033	876	National Railway Museum, Shildon	Gateshead 1889

CLASS J36 0-6-0

Built: 1888–1900. Holmes NBR Class C. 168 built.
Boiler Pressure: 165 lbf/sq in.
Wheel Diameter: 5′ 0″.
Cylinders: 18″ x 26″ (I).
Tractive Effort: 20240 lbf.
RA: 3.
Weight–Loco: 41.95 tons.
–Tender: 33.5 tons.
Valve Gear: Stephenson. Slide valves.
BR Power Classification: 2F.

BR	*LNER*	*NBR*			
65243	9673–5243	673	MAUDE	National Railway Museum, York	N 4392/1891

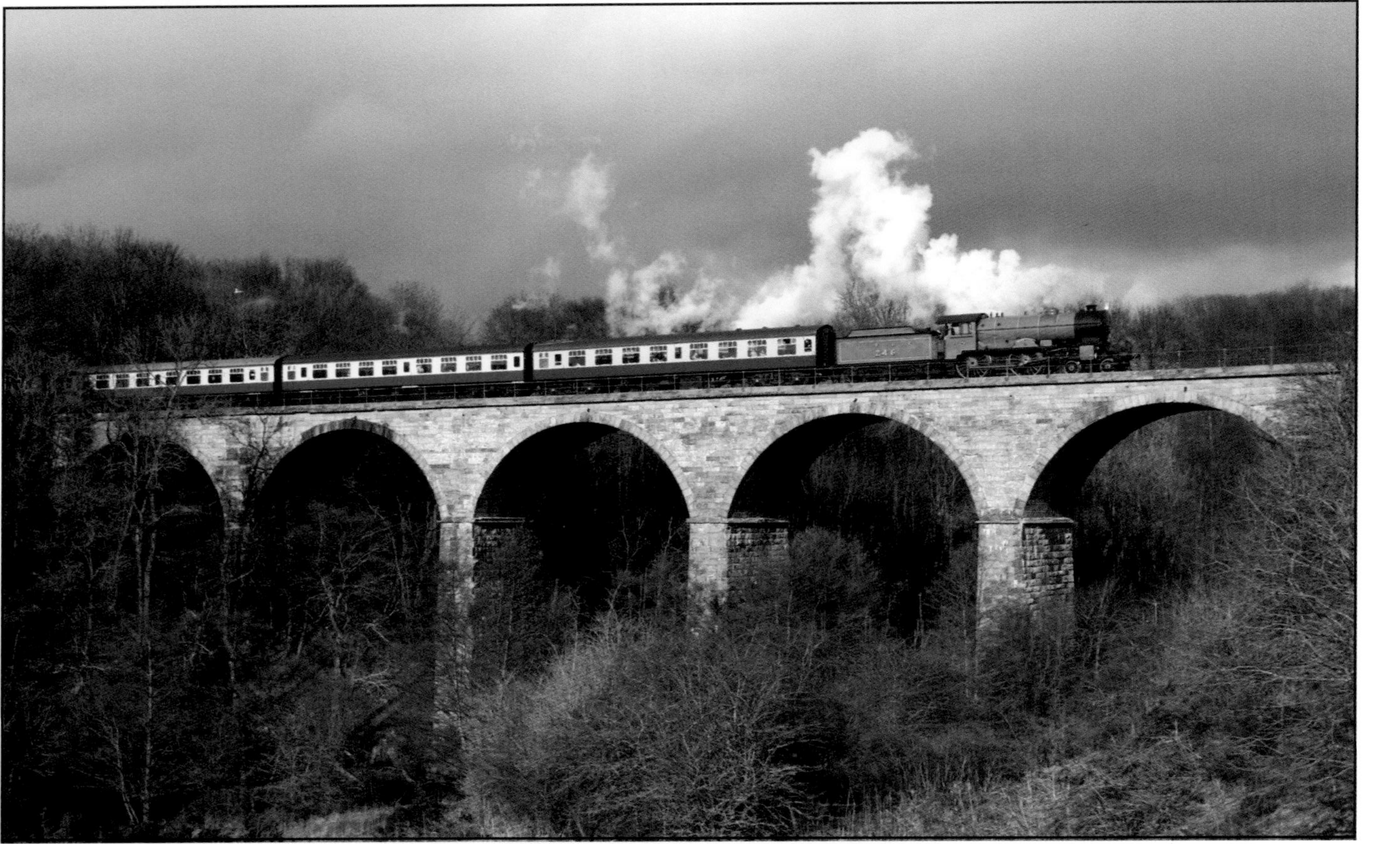

▲ D49/1 4-4-0 246 "MORAYSHIRE" crosses the Avon Viaduct on the Bo'ness & Kinneil Railway extension to Manuel on 27 March 2010 with the 16.30 Bo'ness–Manuel. **Ian Lothian**

▲ North Norfolk Railway-based J15 0-6-0 65462 climbs past Sheringham golf course with the 13.30 Sheringham–Holt on 30 March 2011. **Antony Guppy**

▼ Y7 0-4-0T 985 at the North of England Open Air Museum, Beamish, on 21 May 2011. **Paul Abell**

CLASS J15 0-6-0

Built: 1883–1913. Worsdell GER Class Y14. 189 built.
Boiler Pressure: 160 lbf/sq in.
Wheel Diameter: 4′ 11″.
Cylinders: 17½″ x 24″ (I).
Tractive Effort: 16940 lbf.
RA: 1.
Weight–Loco: 37.1 tons.
–Tender: 30.65 tons.
Valve Gear: Stephenson. Slide valves.
BR Power Classification: 2F.

Dual (air/vacuum) brakes.

BR	*LNER*	*GER*		
65462	7564–5462	564	North Norfolk Railway	Stratford 1912

CLASS J17 0-6-0

Built: 1900–11. Holden GER Class G58. 90 built (65500–89).
Boiler Pressure: 180 lbf/sq in superheated.
Wheel Diameter: 4′ 11″.
Cylinders: 19″ x 26″ (I).
Tractive Effort: 24340 lbf.
RA: 4.
Weight –Loco: 45.4 tons.
–Tender: 38.25 tons.
Valve Gear: Stephenson. Slide valves.
BR Power Classification: 4F.

Air brakes.

BR	*LNER*	*GER*		
65567	8217–5567	1217	Barrow Hill Roundhouse (N)	Stratford 1905

CLASS J27 0-6-0

Built: 1906–23. Worsdell NER Class P3. 115 built.
Boiler Pressure: 180 lbf/sq in superheated.
Wheel Diameter: 4′ 7¼″.
Cylinders: 18½″ x 26″ (I).
Tractive Effort: 24640 lbf.
RA: 5.
Weight–Loco: 47 tons.
–Tender: 37.6 tons.
Valve Gear: Stephenson. Piston valves.
BR Power Classification: 4F.

BR	*LNER*		
65894	2392–5894	NELPG, Hopetown, Darlington	Darlington 1923

CLASS J94. 68077/8 (LNER 8077/8) – see War Department Steam Locomotives.

CLASS Y5 0-4-0ST

Built: 1874–1903. Neilson & Company design for GER (Class 209). 8 built. Survivor sold in 1917.
Boiler Pressure: 140 lbf/sq in.
Wheel Diameter: 3′ 7″.
Valve Gear: Stephenson. Slide valves.
RA: 1.
Weight: 21.2 tons.
Cylinders: 12″ x 20″ (O).
Tractive Effort: 7970 lbf.

GER		
229	Flour Mill Workshop, Bream	N 2119/1876

CLASS Y7 0-4-0T

Built: 1888–1923. Worsdell NER Class H. 24 built.
Boiler Pressure: 160 lbf/sq in.
Wheel Diameter: 4′ 0″.
Cylinders: 14″ x 20″ (I).
BR Power Classification: 0F.
Weight: 22.7 tons.
Tractive Effort: 11140 lbf.
Valve Gear: Joy. Slide valves.
RA: 1.

BR	*LNER*	*NER*		
68088	985–8088		North of England Open Air Museum, Beamish	Darlington 1205/1923
–	1310	1310	Middleton Railway	Gateshead 38/1891

CLASS Y9 — 0-4-0ST

Built: 1882–99. Drummond NBR Class G. 35 built.
Boiler Pressure: 130 lbf/sq in.
Wheel Diameter: 3′ 8″.
Valve Gear: Stephenson. Slide valves.
BR Power Classification: 0F.
Weight: 27.8 tons.
Cylinders: 14″ x 20″ (I).
Tractive Effort: 9840 lbf.
RA: 2.

BR	*LNER*	*NBR*		
68095	10094–8095	42–894–1094	Bo'ness & Kinneil Railway	Cowlairs 1887

CLASS Y1 — 4wT

Built: 1925–33. Sentinel geared loco. 24 built.
Boiler Pressure: 275 lbf/sq in superheated.
Wheel Diameter: 2′ 6″.
Valve Gear: Rotary cam. Poppet valves.
Tractive Effort: 7260 lbf.
RA: 1.
Weight: 19.8 tons.
Cylinders: 6¾″ x 9″ (I).
BR Power Classification: 0F.

BR	*LNER*		
68153	59–8153	Middleton Railway	S 8837/1933

Also carried DEPARTMENTAL LOCOMOTIVE No. 54.

CLASS J69 — 0-6-0T

Built: 1890–1904. Holden GER Class S56. 126 locomotives (including many rebuilt from J67).
Boiler Pressure: 180 lbf/sq in.
Wheel Diameter: 4′ 0″.
Valve Gear: Stephenson. Slide valves.
BR Power Classification: 2F.
Weight: 42.45 tons.
Cylinders: 16½″ x 22″ (I).
Tractive Effort: 19090 lbf.
RA: 3.

Air brakes.

BR	*LNER*	*GER*		
68633	7087–8633	87	National Railway Museum, York (N)	Stratford 1249/1904

CLASS J52 — 0-6-0ST

Built: 1897–1909. Ivatt GNR Class J13. Many rebuilt from Stirling locomotives (built 1892–97).
Boiler Pressure: 170 lbf/sq in.
Wheel Diameter: 4′ 8″.
Valve Gear: Stephenson. Slide valves.
BR Power Classification: 3F.
Weight: 51.7 tons.
Cylinders: 18″ x 26″ (I).
Tractive Effort: 21740 lbf.
RA: 5.

BR	*LNER*	*GNR*		
68846	4247–8846	1247	National Railway Museum, Shildon (N)	SS 4492/1899

CLASS J72 — 0-6-0T

Built: 1898–1925. Worsdell NER Class E1. Further batch built 1949–51 by BR. 113 built.
Boiler Pressure: 140 lbf/sq in.
Wheel Diameter: 4′ 1¼″.
Valve Gear: Stephenson. Slide valves.
BR Power Classification: 2F.
Weight: 38.6 tons.
Cylinders: 17″ x 24″ (I).
Tractive Effort: 16760 lbf.
RA: 5.

69023–Departmental No. 59	National Railway Museum, Shildon	Darlington 2151/1951

CLASS N2 — 0-6-2T

Built: 1920–29. Gresley GNR Class N2. 107 built (69490–69596).
Boiler Pressure: 170 lbf/sq in superheated.
Wheel Diameter: 5′ 8″, 3′ 8″.
Valve Gear: Stephenson. Piston valves.
BR Power Classification: 3MT.
Weight: 70.25 tons.
Cylinders: 19″ x 26″ (I).
Tractive Effort: 19950 lbf.
RA: 6.

BR	*LNER*	*GNR*		
69523	4744–9523	1744	Great Central Railway	NBL 22600/1921

CLASS N7 0-6-2T

Built: 1915–28. Hill GER Class L77. 134 built (69600–69733). 69621 rebuilt to N7/4 1946.
Boiler Pressure: 180 lbf/sq in superheated.
Wheel Diameters: 4′ 10″, 3′ 9″.
Valve Gear: Walschaerts (inside). Piston valves.
RA: 5.
Weight: 61.8 tons.
Cylinders: 18″ x 24″ (I).
Tractive Effort: 20510 lbf.
BR Power Classification: 3MT.

Dual (air/vacuum) brakes.

BR	*LNER*	*GER*			
69621	999E–7999–9621	999	"A.J. HILL"	East Anglian Railway Museum	Stratford 1924

CLASS X1 2-2-4T

Built: 1869 by NER as 2–2–2WT. Rebuilt 1892 to 2-cylinder compound 4–2–2T and rebuilt as 2–2–4T in 1902 and used for pulling inspection saloons. NER Class 66.
Boiler Pressure: 175 lbf/sq in.
Wheel Diameters: 3′ 7″, 5′ 7¾″, 3′ 1¼″.
Valve Gear: Stephenson. Slide valves.
Weight: 44.95 tons.
Cylinders: 13″ x 24″ (hp) + 18½″ x 20″ (lp) (I).
Tractive Effort: 6390 lbf.

LNER	*NER*			
66	1478–66	AEROLITE	National Railway Museum, York (N)	Gateshead 1869

NER 901 CLASS 2-4-0

Built: 1872–82. Fletcher design. 55 built.
Boiler Pressure: 160 lbf/sq in.
Wheel Diameters: 4′ 6″, 7′ 0″.
Cylinders: 18″ x 24″ (I).
Tractive Effort: 12590 lbf.
Weight–Loco: 39.7 tons.
–Tender: 29.9 tons.
Valve Gear: Stephenson. Slide valves.

LNER	*NER*		
910	910	National Railway Museum, Shildon (N)	Gateshead 1875

NER 1001 CLASS 0-6-0

Built: 1864–75. Bouch design for Stockton and Darlington Railway.
Boiler Pressure: 130 lbf/sq in.
Wheel Diameter: 5′ 0½″.
Valve Gear: Stephenson. Slide valves.
Weight–Loco: 35 tons.
Cylinders: 17″ x 26″ (I).
Tractive Effort: 13720 lbf.

LNER	*NER*		
1275	1275	National Railway Museum, York (N)	Glasgow 707/1874

CLASS E5 2-4-0

Built: 1885. Tennant NER 1463 Class. 20 built.
Boiler Pressure: 160 lbf/sq in.
Wheel Diameters: 4′ 6″, 7′ 0″.
Cylinders: 18″ x 24″ (I).
Tractive Effort: 12590 lbf.
Weight–Loco: 42.1 tons.
–Tender: 32.1 tons.
Valve Gear: Stephenson. Slide valves.

LNER	*NER*		
1463	1463	Head of Steam, Darlington Railway Museum (N)	Darlington 1885

CLASS D17/1 4-4-0

Built: 1893–97. Worsdell NER Class M1 (later Class M). 20 built.
Boiler Pressure: 160 lbf/sq in.
Wheel Diameters: 3′ 7¼″, 7′ 1¼″.
Cylinders: 19″ x 26″ (I).
Tractive Effort: 14970 lbf.
Weight–Loco: 52 tons.
–Tender: 41 tons.
Valve Gear: Stephenson. Slide valves.
RA: 6.

LNER	*NER*		
1621	1621	National Railway Museum, York (N)	Gateshead 1893

▲ J72 0-6-0T 69023 passes Swithland Sidings on the Great Central Railway with the 13.00 Loughborough–Leicester North on 10 October 2010. **Hugh Ballantyne**

▼ N7 0-6-2T 69621 at Sheringham on 13 March 2010. **Paul Abell**

CLASS C2 "KLONDYKE" 4-4-2

Built: 1898–1903. HA Ivatt GNR Class C1. 22 built.
Boiler Pressure: 170 lbf/sq in superheated
Wheel Diameters: 3′ 8″, 6′ 8″, 3′ 8″.
Cylinders: 19″ x 24″ (O).
Tractive Effort: 15650 lbf.
Weight–Loco: 62 tons.
–Tender: 42.1 tons.
Valve Gear: Stephenson. Piston valves.
RA: 4.

LNER	*GNR*			
3990	990	HENRY OAKLEY	Bressingham Steam Museum (N)	Doncaster 769/1898

GNR CLASS A2 4-2-2

Built: 1870–93. Stirling design. 47 built.
Boiler Pressure: 140 lbf/sq in.
Wheel Diameters: 3′ 10″, 8′ 1″, 4′ 1″.
Cylinders: 18″ x 28″ (O).
Tractive Effort: 11 130 lbf.
Weight–Loco: 38.5 tons.
–Tender: 30 tons.
Valve Gear: Stephenson. Slide valves.

1 National Railway Museum, Shildon (N) Doncaster 50/1870

STOCKTON & DARLINGTON RAILWAY 0-4-0

Built: 1825–26. George Stephenson design. 6 built.
Boiler Pressure: 50 lbf/sq in.
Wheel Diameter: 3′ 11″.
Cylinders: 9½″ x 24″ (O).
Weight–Loco: 6.5 tons.
–Tender:
Tractive Effort: 2050 lbf.

1 LOCOMOTION Head of Steam, Darlington Railway Museum (N) RS 1/1825

STOCKTON & DARLINGTON RAILWAY 0-6-0

Built: 1845.
Boiler Pressure: 75 lbf/sq in.
Wheel Diameter: 4′ 0″.
Tractive Effort: 6700 lbf.
Weight–Loco: 6.5 tons.
Cylinders: 14½″ x 24″ (O).

25 DERWENT Head of Steam, Darlington Railway Museum (N) Kitching 1845

▶ The only surviving 2-cylinder compound in Britain (a close look will reveal the different diameters of the two cylinders), X1 2-2-4T No. 66 "AEROLITE", was built by the North Eastern Railway in 1869, and rebuilt into its present form in 1902. It makes a rare trip into the NRM York car park on 19 May 2009. **Paul Abell**

1.5. BRITISH RAILWAYS STANDARD STEAM LOCOMOTIVES

GENERAL

From 1951 onwards BR produced a series of standard steam locomotives under the jurisdiction of RA Riddles. Examples of most classes have been preserved, the exceptions being the Class 6MT "Clan" Pacifics, the Class 3MT 2-6-0s (77000 series), the Class 3MT 2-6-2Ts (82000 series) and the Class 2MT 2-6-2Ts (84000 series). Attempts are being made, however, to rectify two of the omissions.

NUMBERING SYSTEM

Tender engines were numbered in the 70000 series and tank engines in the 80000 series, the exceptions being the class 9F 2-10-0s which were numbered in the 92000 series.

CLASSIFICATION SYSTEM

British Railways standard steam locomotives were referred to by power classification like LMS locomotives. All locomotives were classed as "MT" denoting "mixed traffic" except for 71000 and the Class 9F 2-10-0s. The latter, although freight locos, were often used on passenger trains on summer Saturdays.

▲ Britannia 4-6-2 70013 "OLIVER CROMWELL" leaves Loughborough for Leicester North on 26 November 2011. **Hugh Ballantyne**

CLASS 7MT BRITANNIA 4-6-2

Built: 1951–54. 55 built (70000–54).
Boiler Pressure: 250 lbf/sq in superheated.
Wheel Diameters: 3′ 0″, 6′ 2″, 3′ 3½″.
Cylinders: 20″ x 28″ (O).
Tractive Effort: 32 160 lbf.
Weight–Loco: 94 tons.
–Tender: 49.15 tons.
Valve Gear: Walschaerts. Piston valves.
RA: 7

x–Dual (air/vacuum) brakes.

70000 x	BRITANNIA	Southall Depot, London	Crewe 1951
70013	OLIVER CROMWELL	Great Central Railway (N)	Crewe 1951

CLASS 8P 4-6-2

Built: 1954. 1 built.
Boiler Pressure: 250 lbf/sq in superheated.
Wheel Diameters: 3′ 0″, 6′ 2″, 3′ 3½″.
Cylinders: 18″ x 28″ (3).
Tractive Effort: 39 080 lbf.
Weight–Loco: 101.25 tons.
–Tender: 53.7 tons.
Valve Gear: British Caprotti (outside). Poppet valves.
RA: 8.

Dual (air/vacuum) brakes.

71000	DUKE OF GLOUCESTER	East Lancashire Railway	Crewe 1954

CLASS 5MT 4-6-0

Built: 1951–57. 172 built (73000–171).
Boiler Pressure: 225 lbf/sq in superheated.
Wheel Diameters: 3′ 0″, 6′ 2″.
Cylinders: 19″ x 28″ (O).
Valve Gear: Walschaerts. Piston valves (* outside British Caprotti. Poppet valves).
Tractive Effort: 26 120 lbf.
Weight–Loco: 76 tons.
–Tender: 49.15 tons.
RA: 5.

x–Dual (air/vacuum) brakes.

73050 x	"CITY OF PETERBOROUGH"	Nene Valley Railway	Derby 1954
73082	CAMELOT	Bluebell Railway	Derby 1955
73096		Mid Hants Railway	Derby 1955
73129 *		Midland Railway-Butterley	Derby 1956
73156		Great Central Railway	Doncaster 1956

CLASS 4MT 4-6-0

Built: 1951–57. 80 built (75000–79).
Boiler Pressure: 225 lbf/sq in superheated.
Wheel Diameters: 3′ 0″, 5′ 8″.
Cylinders: 18″ x 28″ (O).
Tractive Effort: 25 520 lbf.
Weight–Loco: 67.9 tons.
–Tender: 42.15 tons.
Valve Gear: Walschaerts. Piston valves.
RA: 4.

* Fitted with double chimney.

75014	"BRAVEHEART"	Dartmouth Steam Railway	Swindon 1951
75027		Bluebell Railway	Swindon 1954
75029 *		North Yorkshire Moors Railway	Swindon 1954
75069 *		Severn Valley Railway	Swindon 1955
75078 *		Keighley & Worth Valley Railway	Swindon 1956
75079 *		Mid Hants Railway	Swindon 1956

CLASS 4MT 2-6-0

Built: 1952–57. 115 built (76000–114).
Boiler Pressure: 225 lbf/sq in superheated.
Wheel Diameters: 3′ 0″, 5′ 3″.
Cylinders: 17½″ x 26″ (O).
Tractive Effort: 24 170 lbf.
Weight–Loco: 59.75 tons.
–Tender: 42.15 tons.
Valve Gear: Walschaerts. Piston valves.
RA: 4.

x–Dual (air/vacuum) brakes.

76017	Mid Hants Railway	Horwich 1953
76077	Gloucestershire Warwickshire Railway	Horwich 1956
76079 x	North Yorkshire Moors Railway	Horwich 1957
76084	Ian Storey Engineering, Hepscott	Horwich 1957

CLASS 2MT 2-6-0

Built: 1952–56. 65 built (78000–64). These locomotives were almost identical to the Ivatt LMS Class 2MT 2-6-0s (46400–46527).
Boiler Pressure: 200 lbf/sq in superheated.
Wheel Diameters: 3′ 0″, 5′ 0″.
Cylinders: 16½″ x 24″ (O).
Tractive Effort: 18510 lbf.
Weight–Loco: 49.25 tons.
–Tender: 36.85 tons.
Valve Gear: Walschaerts. Piston valves.
RA: 3.

78018	Darlington North Road Goods Shed	Darlington 1954
78019	Great Central Railway	Darlington 1954
78022	Keighley & Worth Valley Railway	Darlington 1954

CLASS 4MT 2-6-4T

Built: 1951–57. 155 built (80000–154).
Boiler Pressure: 225 lbf/sq in superheated.
Wheel Diameters: 3′ 0″, 5′ 8″, 3′ 0″.
Valve Gear: Walschaerts. Piston valves.
RA: 4.
Weight: 86.65 tons.
Cylinders: 18″ x 28″ (O).
Tractive Effort: 25520 lbf.

80002	Keighley & Worth Valley Railway	Derby 1952
80064	Bluebell Railway	Brighton 1953
80072	Llangollen Railway	Brighton 1953
80078	Swanage Railway	Brighton 1954
80079	Severn Valley Railway	Brighton 1954
80080	Midland Railway-Butterley	Brighton 1954
80097	East Lancashire Railway	Brighton 1954
80098	Midland Railway-Butterley	Brighton 1954
80100	Bluebell Railway	Brighton 1955
80104	Swanage Railway	Brighton 1955
80105	Bo’ness & Kinneil Railway	Brighton 1955
80135	North Yorkshire Moors Railway	Brighton 1956
80136	Crewe Heritage Centre	Brighton 1956
80150	Mid Hants Railway	Brighton 1956
80151	Bluebell Railway	Brighton 1957

CLASS 9F 2-10-0

Built: 1954–60. 251 built (92000–250). 92220 was the last steam locomotive to be built for British Railways.
Boiler Pressure: 250 lbf/sq in superheated.
Wheel Diameters: 3′ 0″, 5′ 0″.
Cylinders: 20″ x 28″ (O).
Tractive Effort: 39670 lbf.
Weight–Loco: 86.7 tons.
–Tender: 52.5 tons.
Valve Gear: Walschaerts. Piston valves.
RA: 9.

* Fitted with single chimney. All other surviving members of the class have double chimneys.

92134 *		Crewe Heritage Centre	Crewe 1957
92203	"BLACK PRINCE"	North Norfolk Railway	Swindon 1959
92207	"MORNING STAR"	North Dorset Railway, Shillingstone	Swindon 1959
92212		Mid Hants Railway	Swindon 1959
92214	"COCK O' THE NORTH"	North Yorkshire Moors Railway	Swindon 1959
92219		Midland Railway-Butterley	Swindon 1960
92220	EVENING STAR	National Railway Museum, York (N)	Swindon 1960
92240		Bluebell Railway	Crewe 1958
92245		Barry Rail Centre	Crewe 1958

▲ Caprotti 5MT 4-6-0 73129 runs through Burrs Country Park shortly after departure from Bury with the 15.20 Heywood–Rawtenstall on 29 January 2011. **Terry Eyres**

▼ 4MT 2-6-0 76079 storms up the 1 in 49 gradient towards Goathland at Darnholm on 4 October 2009 with a morning Grosmont–Pickering train. **Andrew Mason**

▲ Visiting the East Lancashire Railway from the Midland Railway-Butterley, 4MT 2-6-4T 80080 passes Burrs Country Park on a Heywood–Rawtenstall train on 13 March 2011. **Brian Dobbs**

▼ One of the nine preserved 9F 2-10-0s, 92203 "BLACK PRINCE", arrives at Toddington on 19 July 2009, as GWR 4500 Class 2-6-2T 5542 backs onto the stock for the next departure to Cheltenham Racecourse. **John Chalcraft**

1.6. WAR DEPARTMENT LOCOMOTIVES

GENERAL

During the Second World War the War Department (WD) of the British Government acquired and used a considerable number of steam locomotives. On the cessation of hostilities many of these locomotives were sold for further service both to industrial users and other railway administrations. The bulk of WD steam locomotives preserved date from this period. By 1952 many of the large wartime fleet of steam locomotives had been disposed of and those remaining were renumbered into a new series. From 1 April 1964 the WD became the Army Department of the Ministry of Defence with consequent renumbering taking place in 1968. The locomotives considered to be main line locomotives built to "Austerity" designs are included here. Also included in this section are the steam locomotives built to WD "Austerity" designs for industrial users.

CLASS WD AUSTERITY 2-10-0

Built: 1943–45 by North British. 150 built. Many sold to overseas railways. 25 sold to British Railways in 1948 and numbered 90750–74.
Boiler Pressure: 225 lbf/sq in superheated.
Wheel Diameters: 3′ 2″, 4′ 8½″.
Cylinders: 19″ x 28″ (O).
Tractive Effort: 34210 lbf.
Weight–Loco: 78.3 tons.
–Tender: 55.5 tons
Valve Gear: Walschaerts. Piston valves.
BR Power Classification: 8F.

g Hellenic Railways (Greece) number.

WD	*AD/Overseas*			
3651–73651	600	GORDON	Severn Valley Railway	NBL 25437/1943
3652–73652	Lb951 g	"90775"	North Norfolk Railway	NBL 25438/1943
3672–73672	Lb960 g	"DAME VERA LYNN"	North Yorkshire Moors Railway	NBL 25458/1944

CLASS WD AUSTERITY 2-8-0

Built: 1943–45 by North British & Vulcan Foundry. 935 built. Many sold to overseas railways. 200 were sold to LNER in 1946. These became LNER Nos. 3000–3199 and BR 90000–100, 90422–520. A further 533 were sold to British Railways in 1948 and numbered 90101–421, 90521–732.
Boiler Pressure: 225 lbf/sq in superheated.
Wheel Diameters: 3′ 2″, 4′ 8½″.
Cylinders: 19″ x 28″ (O).
Tractive Effort: 34210 lbf.
Weight–Loco: 70.25 tons.
–Tender: 55.5 tons.
Valve Gear: Walschaerts. Piston valves.
BR Power Classification: 8F.

This locomotive was purchased from Swedish State Railways (SJ), it previously having seen service with Netherlands Railways (NS). It has been restored to BR condition.

NS	*SJ*	*WD*	*Present*		
4464	1931	79257	90733	Keighley & Worth Valley Railway	VF 5200/1945

CLASS 50550 0-6-0ST

Built: 1941–42. Eight locomotives were built to this design, all being intended for Stewarts and Lloyd Minerals. Only one was delivered with three being taken over by WD becoming 65–67. The other four went to other industrial users and two of these survive.
Boiler Pressure: 170 lbf/sq in.
Wheel Diameter: 4′ 5″.
Valve Gear: Stephenson. Slide valves.
Weight: 48.35 tons.
Cylinders: 18″ x 26″ (I).
Tractive Effort: 22150 lbf.

WD	*Present*		
–	Unnumbered	Rutland Railway Museum	HE 2411/1941
–	GUNBY	Swindon & Cricklade Railway	HE 2413/1942
66-70066	S112 SPITFIRE	Embsay & Bolton Abbey Railway	HE 2414/1942

CLASS WD AUSTERITY 0-6-0ST

Built: 1943–1953 for Ministry of Supply and War Department. 391 built. 75 bought by LNER and classified J94. Many others passed to industrial users. The design of this class was derived from the "50550" class of Hunslet locomotives (see above).
Boiler Pressure: 170 lbf/sq in.
Wheel Diameter: 4′ 3″.
Valve Gear: Stephenson. Slide valves.
BR Power Classification: 4F.
Weight: 48.25 tons.
Cylinders: 18″ x 26″ (I).
Tractive Effort: 23870 lbf.
RA: 5.

§ Oil fired.
x Dual (air/vacuum) brakes.
† Rebuilt as 0-6-0 tender locomotive. Weights unknown.

BR	*LNER*	*WD*		
68078	8078	71463	Hope Farm, Sellindge	AB 2212/1946
68077	8077	71466	Spa Valley Railway	AB 2215/1946

WD/AD	*Present*		
71480	Unnumbered	Tyseley Locomotive Works, Birmingham	RSH 7289/1945
71499	Unnumbered	Bryn Engineering, Blackrod	HC 1776/1944
71505–118 §	BRUSSELS	Keighley & Worth Valley Railway	HC 1782/1945
71515	MECH NAVIES LTD	Pontypool & Blaenavon Railway	RSH 7169/1944
71516	WELSH GUARDSMAN/ GWARCHODWR CYMREIG	Gwili Railway	RSH 7170/1944
71529–165	No. 15	Rye Farm, Wishaw, Sutton Coldfield	AB 2183/1945
75006	Nene Valley Railway	HE 2855/1943	
75008	SWIFTSURE	Strathspey Railway	HE 2857/1943
75015	48	Strathspey Railway	HE 2864/1943
75019–168	LORD PHIL	Peak Rail	HE 2868/1943 *rebuilt* HE 3883/1962
75030	Unnumbered	Caledonian Railway	HE 2879/1943
75031–101	No. 17	Bo'ness & Kinneil Railway	HE 2880/1943
75041–107†	10 DOUGLAS	Mid Hants Railway	HE 2890/1943 *rebuilt* HE 3882/1962
75050	No. 27 NORMAN	Embsay & Bolton Abbey Railway	RSH 7086/1943
75061	No. 9 CAIRNGORM	Strathspey Railway	RSH 7097/1943
75062	49	Tanfield Railway	RSH 7098/1943
75091	ROBERT	Great Central Railway	HC 1752/1943
75105	WALKDEN	Ribble Steam Railway	HE 3155/1944
75113–132	SAPPER	East Lancashire Railway	HE 3163/1944 *rebuilt* HE 3885/1964
75118–134	S134 WHELDALE	Embsay & Bolton Abbey Railway	HE 3168/1944
75130	No. 3180 ANTWERP	Hope Farm, Sellindge	HE 3180/1944
75133–138	Unnumbered	Flour Mill Workshops, Bream (N)	HE 3183/1944
75141–139 x	68006	Peak Rail	HE 3192/1944 *rebuilt* HE 3888/1964
75142–140	68012 BLACKIE	Mid Norfolk Railway	HE 3193/1944 *rebuilt* HE 3887/1964
75158–144	68012 THE DUKE	Ecclesbourne Valley Railway	WB 2746/1944
75161	Unnumbered	Caledonian Railway	WB 2749/1944
75170	Unnumbered	Cefn Coed Colliery Museum	WB 2758/1944
75171–147	Unnumbered	Caledonian Railway	WB 2759/1944
75178	Unnumbered	Bodmin & Wenford Railway	WB 2766/1945
75186–150	68013	Peak Rail	RSH 7136/1944 *rebuilt* HE 3892/1969
75189–152	No. 8	Flour Mill Workshop, Bream	RSH 7139/1944 *rebuilt* HE 3880/1962
75254–175		Bo'ness & Kinneil Railway	WB 2777/1945
75256	No. 20 TANFIELD	Tanfield Railway	WB 2779/1945
75282–181	HAULWEN	Gwili Railway rebuilt HE 3879/1961	VF 5272/1945
75319	72	Llangollen Railway	VF 5309/1945
190–90		Colne Valley Railway	HE 3790/1952

191–91	No. 23 HOLMAN F. STEPHENS	Kent & East Sussex Railway	HE 3791/1952
192–92	WAGGONER	Isle of Wight Steam Railway	HE 3792/1953
193–93	Unnumbered	Ribble Steam Railway	HE 3793/1953
194–94	CUMBRIA	Ribble Steam Railway	HE 3794/1953
197	No. 25 NORTHIAM	Kent & East Sussex Railway	HE 3797/1953
198–98	ROYAL ENGINEER	Isle of Wight Steam Railway	HE 3798/1953
200	No. 24 ROLVENDEN	Kent & East Sussex Railway	HE 3800/1953

In addition to the 391 locomotives built for the Ministry of Supply and the War Department, a further 93 were built for industrial users, those that survive being:

No. 60	Strathspey Railway	HE 3686/1949
WHISTON	Foxfield Railway	HE 3694/1950
IR	Ribble Steam Railway	HE 3696/1950
11 REPULSE	Lakeside & Haverthwaite Railway	HE 3698/1950
NORMA	Oswestry Railway Centre	HE 3770/1952
8 SIR ROBERT PEEL	Embsay & Bolton Abbey Railway	HE 3776/1952
68030	Strathspey Railway	HE 3777/1952
1	Mid Hants Railway	HE 3781/1952
No. 69	Embsay & Bolton Abbey Railway	HE 3785/1953
N.C.B. MONCKTON No.1	Embsay & Bolton Abbey Railway	HE 3788/1953
WILBERT REV.W.AWDRY	Dean Forest Railway	HE 3806/1953
3809	Great Central Railway	HE 3809/1954
GLENDOWER	South Devon Railway	HE 3810/1954
No. 19	Bo'ness & Kinneil Railway	HE 3818/1954
WARRIOR	Dean Forest Railway	HE 3823/1954
68009	West Coast Railway Company, Carnforth	HE 3825/1954
Unnumbered	Gwili Railway	HE 3829/1955
No. 5	Bo'ness & Kinneil Railway	HE 3837/1955
WIMBLEBURY	Foxfield Railway	HE 3839/1956
PAMELA	Garw Valley Railway	HE 3840/1956
No. 22	Nene Valley Railway	HE 3846/1956
JUNO	National Railway Museum, Shildon	HE 3850/1958
CADLEY HILL No. 1	Snibston Discovery Park	HE 3851/1962
28 – 65	Flour Mill Workshop, Bream	HE 3889/1964
66	Buckinghamshire Railway Centre	HE 3890/1964

SHROPSHIRE & MONTGOMERY RLY 0-4-2WT

Built: 1893 by Alfred Dodman & Company for William Birkitt. Sold to Shropshire & Montgomery Railway in 1911 when it was rebuilt from 2-2-2WT.
Boiler Pressure: 60 lbf/sq in. **Weight:** 5.5 tons.
Wheel Diameters: 2' 3", 2'3". **Cylinders:** 4" x 9" (I).

This locomotive together with three others (LNWR coal engines) became BR (WR) stock in 1950 when this line was nationalised. The locomotives were then withdrawn.

1	GAZELLE	Kent & East Sussex Railway	Dodman 1893

▲ Many preserved lines have at least one Class WD Austerity 0-6-0ST on their books. On 14 August 2010 75282 "HAULWEN" is seen at the Bronwydd Arms base of the Gwili Railway. **Robert Pritchard**

▼ Class S160 2-8-0 (ex-Chinese State Railways) 5197 at Cheddleton on the Churnet Valley Railway on 15 November 2009. **Paul Abell**

1.7. UNITED STATES ARMY TRANSPORTATION CORPS STEAM LOCOMOTIVES

CLASS S160 — 2-8-0

Built: 1942–45 by various American builders. It is estimated that 2120 were built. Many saw use on Britain's railways during the Second World War. Post war many were sold to overseas railway administrations. Only those currently in Great Britain are shown here.
Boiler Pressure: 225 lbf/sq in superheated.
Wheel Diameters: 3′ 2″, 4′ 9″.
Cylinders: 19″ x 26″ (O).
Valve Gear: Walschaerts. Piston valves.
Weight–Loco: 73 tons.
–Tender: 52.1 tons.
Tractive Effort: 31 490 lbf.

USATC	*Overseas Railway*		
1631	MAV (Hungarian Railways) 411.388	Nottingham Transport Heritage Centre	AL 70284/1942
2138	MAV (Hungarian Railways) 411.009	Nottingham Transport Heritage Centre	AL 70620/1943
2253	PKP (Polish Railways) Ty203-288	North Yorkshire Moors Railway	BLW 69496/1943
2364	MAV (Hungarian Railways) 411.337	Nottingham Transport Heritage Centre	BLW 69621/1943
3278	FS (Italian State Railways) 736.073	GCE&SCS, Easton, Isle of Portland	AL 71533/1944
5197	Chinese State Railways KD6.463	Churnet Valley Railway	Lima 8856/1945
5820	PKP (Polish Railways) Ty203.474	Keighley & Worth Valley Railway	Lima 8758/1945
6046	MAV (Hungarian Railways) 411.144	Churnet Valley Railway	BLW 72080/1945

3278 passed to Hellenic Railways (Greece) and numbered 575. Carries name "FRANKLYN D. ROOSEVELT".

See also SR section for USATC 0-6-0Ts.

1.8. NEW BUILD STEAM LOCOMOTIVES

The following is a list of new build projects that have either been completed or are in progress. Technical details refer to the original locos and the new builds may differ from these.

COMPLETED NEW BUILD

LYNTON & BARNSTAPLE 2-6-2T

Currently painted as BR 30190.
Boiler Pressure: 180 lbf/sq in.
Wheel Diameters: 2′ 9″.
Cylinders: 10″ x 16″.
Tractive Effort: 7420 lbf.
Weight: 26 tons.
Valve Gear: Joy

190–30190	LYD	Ffestiniog Railway	Boston Lodge 2010

GWR 4300 CLASS 2-6-0

Original Class built: 1911–32 (see page 17).
Boiler Pressure: 225 lbf/sq in superheated.
Wheel Diameters: 3′ 2″, 5′ 8″.
Cylinders: 18″ x 30 (O).
Tractive Effort: 28880 lbf.
Weight–Loco: 63.85 tons.
–Tender: 40 tons.
Valve Gear: Stephenson. Piston valves.

Rebuilt from 5101 Class 2-6-2T No. 5193 by the West Somerset Railway.

9351	West Somerset Railway	Swindon 1934 rebuilt WSR 2006

LNER CLASS A1 4-6-2

Original Class built: 1948–49. Peppercorn development of Thompson A1/1. None of the class were preserved so this new locomotive has been constructed, working its first services in 2008.
Boiler Pressure: 250 lbf/sq in superheated.
Wheel Diameters: 3′ 2″, 6′ 8″, 3′ 8″.
Cylinders: 19″ x 26″ (3).
Tractive Effort: 37400 lbf.
RA: 9.
Weight–Loco: 104.7 tons.
–Tender: 66.1 tons.
Valve Gear: Walschaerts. Piston valves.
BR Power Classification: 8P.

Dual (air/vacuum) brakes.

60163	TORNADO	National Railway Museum, York	Darlington Hope Street 2195/2008

▲ Replica Manning Wardle 2-6-2T "LYD" approaches Pont Croesor with a train for Portmadog on 11 September 2010. Completed by the Ffestiniog Railway at Boston Lodge in 2010, the loco is currently running as BR 30190. **Chris Parry**

▼ Newly outshopped in BR Brunswick Green, A1 4-6-2 60163 "TORNADO" is seen at Didcot Railway Centre on 11 June 2011. **Alisdair Anderson**

NEW BUILD LOCOS UNDER CONSTRUCTION

There are two Great Western designs which are being produced using parts from other locos to a greater or lesser extent. These are detailed here.

GWR design

2900 CLASS SAINT 4-6-0

Original Class Built: 1902–13. 77 built (2910–55, 2971–90, 2998). The new build loco is rebuilt from 4942 MAINDY HALL.
Boiler Pressure: 225 lbf/sq in superheated.
Wheel Diameters: 3′ 2″, 6′ 8½″.
Cylinders: 18½″ x 30″ (O).
Tractive Effort: 24390 lbf.
Power Classification: D (4P).
Weight–Loco: 72.0 tons.
–Tender: 43.15 tons.
Valve Gear: Stephenson.
Restriction: Red.

2999	LADY OF LEGEND	Didcot Railway Centre	Under construction

6800 CLASS GRANGE 4-6-0

Original Class Built: 1936–39. 80 built (6800–79). The new build loco uses the boiler of 7927 WILLINGTON HALL.
Boiler Pressure: 225 lbf/sq in superheated.
Wheel Diameters: 3′ 0″, 5′ 8″.
Cylinders: 18½″ x 30″ (O).
Tractive Effort: 28875 lbf.
Power Class: D (5MT).
Weight–Loco: 74.00 tons.
–Tender: 40.00 tons.
Valve Gear: Stephenson. Piston valves.
Restriction: Red.

6880	BETTON GRANGE	Llangollen Railway	Under construction

British Railways design

CLASS 6MT CLAN 4-6-2

Ten Clans were built, numbered 72000–09 in 1951–52 and were all given names of Scottish Clans. Another 15 were ordered, but the order was cancelled and they were never built. The first five of these were intended for the Southern Region and were to be named HENGIST, HORSA, CANUTE, WILDFIRE and FIREBRAND. The new locomotive will therefore be known as 72010 HENGIST.
Boiler Pressure: 225 lbf/sq in (superheated).
Wheel Diameters: 3′ 0″, 6′ 2″, 3′ 3½″.
Cylinders: 19½″ x 28″ (O).
Tractive Effort: 27 520 lbf.
Weight–Loco: 88.5 tons.
–Tender: 49.15 tons.
Valve Gear: Walschaerts. Piston valves.
RA: 6

72010	HENGIST	North Dorset Railway, Shillingstone	Under construction

CLASS 3MT 2-6-2T

This class numbered 45 locos (82000–44) and were built 1952–55.
Boiler Pressure: 200 lbf/sq in (superheated).
Wheel Diameters: 3′ 0″, 5′ 3″, 3′ 0″.
Valve Gear: Walschaerts. Piston valves.
RA: 4.
Weight: 74.05 tons.
Cylinders: 17½″ x 26″ (O).
Tractive Effort: 21 490 lbf.

82045	Severn Valley Railway	Under construction

CLASS 2MT 2-6-2T

This locomotive was built in 1956 as 2MT 2-6-0 78059. The Bluebell Railway has the engine only, and is converting it to a BR 2MT 2-6-2T since none of those survived into preservation.
Boiler Pressure: 200 lbf/sq in (superheated).
Wheel Diameters: 3′ 0″, 5′ 0″, 3′ 0″.
Valve Gear: Walschaerts. Piston valves.
RA: 1.
Weight: 66.25 tons.
Cylinders: 16½″ x 24″ (O).
Tractive Effort: 18 510 lbf.

84030	Bluebell Railway	Darlington 1956 reb. Bluebell

1.9. REPLICA STEAM LOCOMOTIVES

GENERAL

Details are included of locomotives which would have been included in the foregoing had the original locomotive survived. Locomotives are listed under the heading of the original Railway/Manufacturer. Only locomotives which work, previously worked or can be made to work are included.

STOCKTON & DARLINGTON RAILWAY 0-4-0

Built: 1975.
Boiler Pressure: 50 lbf/sq in.
Wheel Diameter: 3′ 11″.
Cylinders: 9½″ x 24″ (O).
Weight–Loco: 6.5 tons.
–Tender:
Tractive Effort: 1960 lbf.

LOCOMOTION	North of England Open Air Museum, Beamish	Loco. Ent.No. 1/1975

LIVERPOOL & MANCHESTER RAILWAY 0-2-2

Built: 1934/79.
Boiler Pressure: 50 lbf/sq in.
Wheel Diameters: 4′ 8½″, 2′ 6″.
Cylinders: 8″ x 17″ (O).
Weight–Loco: 4.5 tons.
–Tender: 5.2 tons.
Tractive Effort: 820 lbf.

ROCKET	National Railway Museum, Shildon (N)	RS 4089/1934
ROCKET	National Railway Museum, York (N)	Loco. Ent.No. 2/1979

LIVERPOOL & MANCHESTER RAILWAY 0-4-0

Built: 1979.
Boiler Pressure: 50 lbf/sq in.
Wheel Diameter: 4′ 6″.
Cylinders: 7″ x 18″ (O).
Weight–Loco: 4.5 tons.
–Tender: 5.2 tons.
Tractive Effort: 690 lbf.

SANS PAREIL	National Railway Museum, Shildon	Shildon/1980

▲ The 1975 replica of 0-4-0 "LOCOMOTION" sits in the sunshine at the North of England Open Air Museum at Beamish on 3 September 2010. **Paul Abell**

LIVERPOOL & MANCHESTER RAILWAY 2-2-0

Built: 1992 (original dated from 1832).
Boiler Pressure: 100 lbf/sq in (original 50 lbf/sq in).
Wheel Diameters: 5′.
Cylinders: 9″ x 22″ (O).
Weight–Loco: 8 tons.
–Tender: 4 tons.
Tractive Effort:

PLANET 9 Museum of Science & Industry, Manchester Manch 1992

BRAITHWAITE & ERICSSON & COMPANY 0-2-2WT

Built: 1929 (using parts from 1829 original)/1980.
Boiler Pressure:
Wheel Diameter:
Cylinders:
Weight–Loco:
–Tender:
Tractive Effort:

NOVELTY Museum of Science & Industry, Manchester (N) Science Museum 1929
NOVELTY Swedish Railway Museum, Gayle, Sweden Loco. Ent. No. 3 1980

GREAT WESTERN RAILWAY 2-2-2

Built: 1925 (using parts from 1837 original).
Boiler Pressure: 90 lbf/sq in.
Wheel Diameters: 4′ 0″, 7′ 0″,4′ 0″.
Cylinders: 16″ x 16″ (I).
Gauge: 7′ 0¼″.
Weight–Loco: 23.35 tons.
–Tender: 6.5 tons.
Tractive Effort: 3730 lbf.

NORTH STAR Steam – Museum of the Great Western Railway (N) Swindon 1925

GREAT WESTERN RAILWAY 2-2-2

Built: 2005.
Boiler Pressure: 50 lbf/sq in.
Wheel Diameters: 4′ 0″, 7′ 0″, 4′ 0″.
Cylinders: 15″ x 18″ (I).
Gauge: 7′ 0¼″.
Weight–Loco: 23.35 tons.
–Tender: 6.5 tons.
Tractive Effort: 2050 lbf.

FIRE FLY Didcot Railway Centre Didcot/Bristol 2005

GREAT WESTERN RAILWAY 4-2-2

Built: 1985.
Boiler Pressure:
Wheel Diameters: 4′ 6″, 8′ 0″, 4′ 6″.
Cylinders: 18″ x 24″ (I).
Gauge: 7′ 0¼″.
Weight–Loco: 35.5 tons.
–Tender:
Tractive Effort: 7920 lbf.

IRON DUKE Gloucestershire Warwickshire Railway (N) Resco 1985

REPLICA LOCOS UNDER CONSTRUCTION

GWR design

1000 CLASS COUNTY 4-6-0

Original Class Built: 1945–47. 30 built (1000–29). The replica loco uses the frames from 7927 WILLINGTON HALL and assumes the identity of a former member of the class. The boiler is rebuilt from Stanier 8F 2-8-0 48518.
Boiler Pressure: 250 lbf/sq in superheated.
Wheel Diameters: 3′ 0″, 6′3″.
Cylinders: 18½″ x 30″ (O).
Tractive Effort: 29090 lbf.
Power Classification: D (6MT).
Weight–Loco: 76.85 tons.
–Tender: 49.00 tons.
Valve Gear: Stephenson. Piston valves.
Restriction: Red.

1014 COUNTY OF GLAMORGAN Didcot Railway Centre Under construction

Southern Railway design

CLASS H2 4-4-2

Original Class Built: 1911–12. 6 built (32421–32426). The replica loco uses the boiler from a GNR C1 Atlantic which is similar to the LBSCR class.

Boiler Pressure: 170 lbf/sq in.
Wheel Diameters: ?", 6'7½".
Cylinders: 21" x 26" (O).
Tractive Effort: 20840 lbf.
Weight–Loco: 68.25 tons.
–Tender: 39.25 tons.
Valve Gear: Stephenson. Piston valves.
BR Power Classification: 3P.

32424	BEACHY HEAD	Bluebell Railway	Under construction

LMS design

CLASS 6P (Formerly 5XP) PATRIOT 4-6-0

Original Class built: 1930–34. Fowler design. The first two members of the class were rebuilt from LNWR "Claughtons". 52 built (45500–45551). This will be the 53rd loco built but will assume the identity of a former member of the class, 45551.

Boiler Pressure: 200 lbf/sq in. superheated.
Wheel Diameters: 3' 3", 6' 9".
Cylinders: 18" x 26" (3).
Tractive Effort: 26520 lbf.
Weight–Loco: 80.75 tons.
–Tender: 42.70 tons.
Valve Gear: Walschaerts. Piston valves.

BR	*LMS*			
45551	5551	THE UNKNOWN WARRIOR	Llangollen Railway	Under construction

LNER design. A third scheme still at an earlier stage of development is for a new build B17 4-6-0 "Sandringham".

CLASS F5 2-4-2T

Original Class built: 1903–09. Holden Great Eastern design for suburban use. The loco is eventually destined for the Epping–Ongar line. The original F5s finished at 67217, but two other locos, 67218/19 were reclassified to F5 from F6 in 1948.

Boiler Pressure: 180 lbf/sq in.
Wheel Diameters: 3' 9", 5' 4", 3' 9".
Valve Gear: Stephenson. Slide valves.
BR Power Classification: 1P.
Weight: 53.95 tons.
Cylinders: 17" x 24" (I).
Tractive Effort: 17570 lbf.
RA: 3.

67218	Mangapps Railway Museum	Under construction

CLASS G5 0-4-4T

Original Class Built: 1894–1901. Wilson Worsdell North Eastern Railway design for branch line and suburban use. 110 built (67240–67349). LNER 7306 was one of two members of the class withdrawn in 1948 and never received its allocated BR number 67306. The original was built at Darlington in 1897 as NER 1759.

Boiler Pressure: 160 lbf/sq in.
Wheel Diameters: 3' 1¼", 5' 1¼".
Valve Gear: Stephenson. Slide valves.
BR Power Classification: 1P.
Weight: 54.20 tons.
Cylinders: 18" x 24" (I).
Tractive Effort: 17200 lbf.
RA: 4.

67306	National Railway Museum, Shildon	Under construction

2. DIESEL LOCOMOTIVES

GENERAL

It was not until the mid-1950s that diesel locomotives appeared in great numbers. However, during the 1930s, particularly on the LMS with diesel shunting locomotives, several small building programmes were authorised. A few locomotives survive from this period along with several others built in the immediate post-war period. During World War II a large proportion of LMS locomotives were transferred to the War Department who also authorised the construction of further orders. Many of these were shipped across the English Channel and were subsequently lost in action. It is, however, possible some still remain undiscovered by the authors. Notification of these would be gratefully appreciated. Those known to survive are listed in Section 2.1.

From the mid-1950s the re-equipment of the railway network began in earnest and vast numbers of new, mainly diesel locomotives were constructed by British Railways and private contractors. Those from this period now considered to be preserved can be found in Section 2.3. Some examples from this period are still in service with main line and industrial users, details of these can be found in the Platform 5 Locomotives pocket books or the bound volume "Locomotives & Coaching Stock".

NOTES

WHEEL ARRANGEMENT

The Whyte notation is used for diesel shunters with coupled driving wheels (see steam section). For other shunting and main line diesel and electric locomotives, the system whereby the number of powered axles on a bogie or frame is denoted by a letter (A = 1, B = 2, C = 3 etc) and the number of unpowered axles is denoted by a number is used. The letter "o" after a letter indicates that each axle is individually powered and a + sign indicates that the bogies are intercoupled.

DIMENSIONS

SI units are generally used. Imperial units are sometimes given in parentheses.

TRACTIVE EFFORT

Continuous and maximum tractive efforts are generally quoted for vehicles with electric transmission.

BRAKES

Locomotives are assumed to have train vacuum brakes unless otherwise stated.

NUMBERING SYSTEMS

Prior to nationalisation each railway company allocated locomotive numbers in accordance with its own policy. However, after nationalisation in 1948 a common system was devised and internal combustion locomotives were allocated five figure numbers in the series 10000–19999.

Diesel locomotives built prior to nationalisation or to pre-nationalisation designs are arranged generally in order of the 1948 numbers with those withdrawn before 1948 listed at the beginning of each section. In 1957 a new numbering scheme was introduced for locomotives built to British Railways specifications and details of this are given in the introduction to the British Railways section.

2.1. LONDON MIDLAND & SCOTTISH RAILWAY

DIESEL MECHANICAL 0-4-0

Built: 1934 by English Electric at Preston Works for Drewry Car Company.
Engine: Allan 8RS18 of 119 kW (160 hp) at 1200 rpm (Now fitted with a Gardner 6L3 of 114 kW (153 hp)).
Transmission: Mechanical. Wilson four-speed gearbox driving a rear jackshaft.
Maximum Tractive Effort: 50 kN (11 200 lbf). **Weight:** 25.8 tonnes.
Maximum Speed: 12 mph. **Wheel Diameter:** 914 mm.

No train brakes.

LMS	*WD*	*AD*		
7050	224–70224–846	240	National Railway Museum, York (N)	DC 2047/EE/DK 847 1934

DIESEL MECHANICAL 0-6-0

Built: 1932 by Hunslet Engine Company (taken into stock 1933).
Engine: MAN 112 kW (150 hp) at 900 rpm (Now fitted with a Maclaren/Ricardo 98 kW engine).
Transmission: Mechanical. Hunslet constant mesh four-speed gearbox.
Maximum Tractive Effort: **Weight:** 21.7 tonnes.
Maximum Speed: 30 mph. **Wheel Diameter:** 914 mm.

Built without train brakes, but vacuum brakes now fitted.

7401–7051	"JOHN ALCOCK"	Middleton Railway	HE 1697/1933

DIESEL ELECTRIC 0-6-0

Built: 1935 by English Electric. 11 built, 3 taken over by BR as 12000–02 (LMS 7074/76/79). Scrapped 1956–62. Others sold to WD for use in France.
Engine: English Electric 6K of 261 kW (350 hp) at 675 rpm.
Transmission: Electric. Two axle-hung traction motors with a single reduction drive.
Maximum Tractive Effort: 147 kN (33 000 lbf). **Weight:** 52 tonnes.
Maximum Speed: 30 mph. **Wheel Diameter:** 1232 mm.

No train brakes.

LMS	*WD*		
7069	18	Gloucestershire Warwickshire Railway	EE/HL 3841 1935

DIESEL ELECTRIC 0-6-0

Built: 1939–42 at Derby. 40 built, 30 taken over by BR as 12003–32 (LMS 7080–99, 7110–19). Scrapped 1964–67. Others sold to WD for use in Italy and Egypt.
Engine: English Electric 6KT of 261 kW (350 hp) at 680 rpm.
Transmission: Electric. One traction motor with jackshaft drive.
Maximum Tractive Effort: 147 kN (33 000 lbf). **Weight:** 56 tonnes.
Maximum Speed: 20 mph. **Wheel Diameter:** 1295 mm.

Built without train brakes, but air brakes now fitted.

LMS	*WD*	*FS*	*Present*			
7103	52–70052		700.001	Unnumbered	Piedmont Rail Museum, Torino, Italy	Derby 1941
7106	55–70055	700.003		Unnumbered	LFI, Arezzostia Pescaiola, Italy	Derby 1941

BR CLASS 11 DIESEL ELECTRIC 0-6-0

Built: 1945–52. LMS design. 120 built. The first order for 20 was for the WD, 14 being delivered as 260–268/70269–273, the balance passing to the LMS as 7120–7125 (BR 12033–038). 260–268 were renumbered 70260–268 and 70260–269 were sold to the NS (Netherlands Railways) in 1946.
Engine: English Electric 6KT of 261 kW (350 hp) at 680 r.p.m.
Transmission: Electric. Two EE 506 axle hung traction motors.
Maximum Tractive Effort: 156 kN (35 000 lbf).
Continuous Tractive Effort: 49.4 kN (11 100 lbf) at 8.8 mph.
Wheel Diameter: 1372 mm. **Weight:** 56 tonnes.
Power at Rail: 183 kW (245 hp). **Maximum Speed:** 20 mph.

No train brakes except 70272, 12099 and 12131 which have since been fitted with vacuum brakes and 12083 which has since been fitted with air brakes.

BR	*WD*	*AD*	*Present*		
–	70269		508§	Netherlands National Railway Museum, Blerik Yard Storage Shed	Derby 1944
–	70272–878	601	7120	Lakeside & Haverthwaite Railway	Derby 1944
12052			MP228	Caledonian Railway	Derby 1949
12061				Peak Rail	Derby 1949
12077				Midland Railway-Butterley	Derby 1950
12082			12049	Mid Hants Railway	Derby 1950
12083				Battlefield Railway	Derby 1950
12093			MP229	Caledonian Railway	Derby 1951
12099				Severn Valley Railway	Derby 1952
12131				North Norfolk Railway	Darlington 1952

§ NS number.

2.2. SOUTHERN RAILWAY

BR CLASS 12 DIESEL ELECTRIC 0-6-0

Built: 1949–52. Based on pre-war LMS design. 26 built.
Engine: English Electric 6KT of 350 h.p. at 680 r.p.m.
Transmission: Electric. Two EE 506A axle-hung traction motors.
Maximum Tractive Effort: 107 kN (24 000 lbf).
Continuous Tractive Effort: 36 kN (8000 lbf) at 10.2 mph.
Wheel Diameter: 1370 mm. **Weight:** 45 tonnes.
Power at Rail: 163 kW (218 hp). **Maximum Speed:** 27 mph.

No train brakes.

15224	Spa Valley Railway	Ashford 1949

2.3. BRITISH RAILWAYS

NUMBERING SYSTEM

In 1957 British Railways introduced a new numbering system which applied to all diesel locomotives except those built to pre-nationalisation designs. Each locomotive was allocated a number of up to four digits prefixed with a "D". Diesel electric shunters already built numbered in the 13xxx series had the "1" replaced by a "D". Diesel mechanical shunters already built numbered in the 11xxx series were allocated numbers in the D2xxx series.

When all steam locomotives had been withdrawn, the prefix letter was officially eliminated from the number of diesel locomotives, although it continued to be carried on many of them. For this reason, no attempt is made to distinguish between those locomotives which did or did not have the "D" prefix removed. Similarly, in preservation, no distinction is made between locomotives which do or do not carry a "D" prefix at present. In 1968 British Railways introduced a numerical classification code for diesel & electric locomotives.

With the introduction of modern communications each locomotive was allocated a two digit class number followed by a three digit serial number. These started to be applied in 1972 and several locomotives have carried more than one number in this scheme.

Locomotives in this section are listed in 1957 number order within each class. A number of diesel shunting locomotives were withdrawn prior to the classification system being introduced and are listed at the end of this section. Experimental and civil engineers' main line locomotives are listed in Sections 2.4 and 2.5.

CLASSIFICATION SYSTEM

It was not until the British Railways organisation was set up that some semblance of order was introduced. This broadly took the following form:

Type	*Engine Horsepower (hp)*	*Number Range*
1	800–1000	D 8000–D 8999
2	1001–1499	D 5000–D 6499
3	1500–1999	D 6500–D 7999
4	2000–2999	D 1 –D 1999
5	3000+	D 9000–D 9499
Shunting	150/300	D 2000–D 2999
Shunting	350/400	D 3000–D 4999
Shunting/trip	650	D 9500–D 9999
AC Electric		E 1000–E 4999
DC Electric		E 5000–E 6999

▶ The Heritage Shunter Trust based at Rowsley on Peak Rail has a fine collection of preserved shunters and sometimes hosts shunter galas. One such event on 27 September 2009 featured this line up of 0-4-0 Class 02 D2854, and 0-6-0s PWM 654 and 03099. **Alisdair Anderson**

▶ BR Blue 0-6-0 03066 at Barrow Hill Roundhouse on 9 October 2010. **Paul Abell**

◀ In contrast, Class 04 0-6-0 D2334 carries original BR Green when seen at Kingsley & Froghall on 15 November 2009. **Paul Abell**

CLASS 01 — 0-4-0

Built: 1956 by Andrew Barclay, Kilmarnock. 4 built.
Engine: Gardner 6L3 of 114 kW (153 hp) at 1200 rpm.
Transmission: Mechanical. Wilson SE4 epicyclic gearbox.
Maximum Tractive Effort: 56.8 kN (12 750 lbf). **Weight**: 25.5 tonnes.
Wheel Diameter: 965 mm. **Maximum Speed**: 14 mph.

No train brakes.

11503–D 2953	Peak Rail	AB 395/1956
11506–D 2956	East Lancashire Railway	AB 398/1956

CLASS 02 — 0-4-0

Built: 1960–61 by Yorkshire Engine Company, Sheffield. 20 built.
Engine: Rolls Royce C6NFL of 127 kW (170 hp) at 1800 rpm.
Transmission: Hydraulic. Rolls Royce CF 10000.
Maximum Tractive Effort: 66.8 kN (15 000 lbf).
Continuous Tractive Effort: 61 kN (13 700 lbf) at 1.4 mph.
Weight: 28.6 tonnes.
Wheel Diameter: 1067 mm. **Maximum Speed**: 30 mph.

D 2853–02003	Appleby-Frodingham RPS, Scunthorpe	YE 2812/1960
D 2854	Peak Rail	YE 2813/1960
D 2858	Midland Railway-Butterley	YE 2817/1960
D 2860	National Railway Museum, York (N)	YE 2843/1961
D 2866	Peak Rail	YE 2849/1961
D 2867	Battlefield Railway	YE 2850/1961
D 2868	Peak Rail	YE 2851/1961

CLASS 03 — 0-6-0

Built: 1957–62 at Doncaster and Swindon. 230 built.
Engine: Gardner 8L3 of 152 kW (204 hp) at 1200 rpm. Replaced with VM V12 (300–350 hp) on D 2128 and D 2134.
Transmission: Mechanical. Wilson CA5 epicyclic gearbox.
Maximum Tractive Effort: 68 kN (15 300 lbf). **Weight**: 31 tonnes.
Wheel Diameter: 1092 mm. **Maximum Speed**: 28 mph.

x Dual braked (air/vacuum). a Air braked only.

§ modified with cut down cab for working on Burry Port & Gwendraeth Valley Line.

11205–D 2018–03018		Mangapps Railway Museum	Swindon 1958
11207–D 2020–03020		Mangapps Railway Museum	Swindon 1958
11209–D 2022–03022		Swindon & Cricklade Railway	Swindon 1958
11210–D 2023		Kent & East Sussex Railway	Swindon 1958
11211–D 2024	No. 4	Kent & East Sussex Railway	Swindon 1958
D 2027–03027		Peak Rail	Swindon 1958
D 2037–03037		Peak Rail	Swindon 1959
D 2041		Colne Valley Railway	Swindon 1959
D 2046		Plym Valley Railway	Doncaster 1958
D 2051	Unnumbered	North Norfolk Railway	Doncaster 1959
D 2059–03059x		Isle of Wight Steam Railway	Doncaster 1959
D 2062–03062		East Lancashire Railway	Doncaster 1959
D 2063–03063x		North Norfolk Railway	Doncaster 1959
D 2066–03066x		Barrow Hill Roundhouse	Doncaster 1959
D 2069–03069		Gloucestershire Warwickshire Railway	Doncaster 1959
D 2072–03072		Lakeside & Haverthwaite Railway	Doncaster 1959
D 2073–03073x		Crewe Heritage Centre	Doncaster 1959
D 2078–03078x		Stephenson Railway Museum	Doncaster 1959
D 2079–03079		Derwent Valley Light Railway	Doncaster 1960
D 2081–03081		Mangapps Railway Museum	Doncaster 1960

D 2084–03084x	"HELEN-LOUISE"	West Coast Railway Company, Carnforth	Doncaster 1959
D 2089–03089x		Mangapps Railway Museum	Doncaster 1960
D 2090–03090		National Railway Museum, Shildon (N)	Doncaster 1960
D 2094–03094x		Royal Deeside Railway	Doncaster 1960
D 2099–03099		Peak Rail	Doncaster 1960
D 2113–03113		Peak Rail	Doncaster 1960
D 2117		Lakeside & Haverthwaite Railway	Swindon 1959
D 2118		Nottingham Transport Heritage Centre	Swindon 1959
D 2119–03119§		West Somerset Railway	Swindon 1959
D 2120–03120§		Fawley Hill Railway	Swindon 1959
D 2128–03128a		Appleby-Frodingham RPS, Scunthorpe	Swindon 1960
D 2133		West Somerset Railway	Swindon 1960
D 2134–03134a		Royal Deeside Railway	Swindon 1960
D 2138		Midland Railway-Butterley	Swindon 1960
D 2139	D2000	Peak Rail	Swindon 1960
D 2141–03141§		Pontypool & Blaenavon Railway	Swindon 1960
D 2144–03144§x	"WESTERN WAGGONER"	Wensleydale Railway	Swindon 1960
D 2145–03145§		Moreton Park Railway	Swindon 1960
D 2148		Ribble Steam Railway	Swindon 1960
D 2152–03152§		Swindon & Cricklade Railway	Swindon 1960
D 2158–03158x	"MARGARET-ANN"	Great Central Railway	Swindon 1960
D 2162–03162x		Llangollen Railway	Swindon 1960
D 2170–03170x		Epping–Ongar Railway	Swindon 1960
D 2178		Gwili Railway	Swindon 1962
D 2180–03180x		Battlefield Railway	Swindon 1962
D 2182		Gloucestershire Warwickshire Railway	Swindon 1962
D 2184		Colne Valley Railway	Swindon 1962
D 2189–03189		Ribble Steam Railway	Swindon 1961
D 2192	"TITAN"	Dartmouth Steam Railway	Swindon 1961
D 2197–03197x		Mangapps Railway Museum	Swindon 1961
D 2199		Peak Rail	Swindon 1961
Dept 92–D 2371–03371x		Rowden Mill Station, Herefordshire	Swindon 1958
D 2399–03399x		Mangapps Railway Museum	Doncaster 1961

03144 is on loan from MoD 275 Squadron RLC (V).

CLASS 04 — 0-6-0

Built: 1952–62. Drewry design built by Vulcan Foundry & Robert Stephenson & Hawthorns. 140 built.
Engine: Gardner 8L3 of 152 kW (204 hp) at 1200 rpm.
Transmission: Mechanical. Wilson CA5 epicyclic gearbox.
Maximum Tractive Effort: 69.7 kN (15 650 lbf). **Weight:** 32 tonnes.
Wheel Diameter: 1067 mm. **Maximum Speed:** 28 mph.

11103–D 2203		Embsay & Bolton Abbey Railway	DC/VF 2400/D145/1952
11106–D 2205		West Somerset Railway	DC/VF 2486/D212/1953
11108–D 2207		North Yorkshire Moors Railway	DC/VF 2482/D208/1953
11135–D 2229	Unnumbered	Peak Rail	DC/VF 2552/D278/1955
11215–D 2245		Battlefield Railway	DC/RSH 2577/7864/1956
11216–D 2246		South Devon Railway	DC/RSH 2578/7865/1956
D 2271		West Somerset Railway	DC/RSH 2615/7913/1958
D 2272		Peak Rail	DC/RSH 2616/7914/1958
D 2279		East Anglian Railway Museum	DC/RSH 2656/8097/1960
D 2280	Unnumbered	North Norfolk Railway	DC/RSH 2657/8098/1960
D 2284		Peak Rail	DC/RSH 2661/8102/1960
D 2298		Buckinghamshire Railway Centre	DC/RSH 2679/8157/1960
D 2302		Barrow Hill Roundhouse	DC/RSH 2683/8161/1960
D 2310	04110	Battlefield Railway	DC/RSH 2691/8169/1960
D 2324		Peak Rail	DC/RSH 2705/8183/1961
D 2325		Mangapps Railway Museum	DC/RSH 2706/8184/1961
D 2334		Churnet Valley Railway	DC/RSH 2715/8193/1961
D 2337		Peak Rail	DC/RSH 2718/8196/1961

CLASS 05 — 0-6-0

Built: 1955–61 by Hunslet Engine Company, Leeds. 69 built.
Engine: Gardner 8L3 of 152 kW (204 hp) at 1200 rpm.
Transmission: Mechanical. Hunslet gearbox.
Maximum Tractive Effort: 64.6 kN (14 500 lbf). **Weight:** 31 tonnes.
Wheel Diameter: 1121 (* 1016) mm. **Maximum Speed:** 18 mph.

11140–D 2554–05001–97803*	Isle of Wight Steam Railway	HE 4870/1956
D 2578	Moreton Park Railway	HE 5460/1958 rebuilt HE 6999/1968
D 2587	Peak Rail	HE 5636/1959 rebuilt HE 7180/1969
D 2595	Ribble Steam Railway	HE 5644/1960 rebuilt HE 7179/1969

CLASS 06 — 0-4-0

Built: 1958–60 by Andrew Barclay, Kilmarnock. 35 built.
Engine: Gardner 8L3 of 152 kW (204 hp) at 1200 rpm.
Transmission: Mechanical. Wilson CA5 epicyclic gearbox.
Maximum Tractive Effort: 88 kN (19 800 lbf). **Weight:** 37 tonnes.
Wheel Diameter: 1092 mm. **Maximum Speed:** 23 mph.

D 2420–06003–97804	Peak Rail	AB 435/1959

CLASS 07 — 0-6-0

Built: 1962 by Ruston & Hornsby, Lincoln. 14 built.
Engine: Paxman 6RPHL Mk III of 205 kW (275 hp) at 1360 rpm.
Transmission: Electric. One AEI RTB 6652 traction motor.
Maximum Tractive Effort: 126 kN (28 240 lbf).
Continuous Tractive Effort: 71 kN (15 950 lbf) at 4.38 mph.
Power at Rail: 142 kW (190 hp). **Weight:** 43.6 tonnes.
Wheel Diameter: 1067 mm. **Maximum Speed:** 20 mph.

x Dual (air/vacuum) braked.

D 2989–07005x	Great Central Railway	RH 480690/1962
D 2994–07010	Avon Valley Railway	RH 480695/1962
D 2995–07011x	St Leonards Railway Engineering	RH 480696/1962
D 2996–07012	Appleby-Frodingham RPS, Scunthorpe	RH 480697/1962
D 2997–07013x	Peak Rail	RH 480698/1962

CLASS 08 — 0-6-0

Built: 1952–62. Built at Derby, Darlington, Crewe, Horwich & Doncaster. 996 built.
Engine: English Electric 6KT of 298 kW (400 hp) at 680 rpm.
Transmission: Electric. Two EE 506 axle-hung traction motors.
Maximum Tractive Effort: 156 kN (35 000 lbf).
Continuous Tractive Effort: 49.4 kN (11 100 lbf) at 8.8 mph.
Power at Rail: 194 kW (260 hp). **Weight:** 50 tonnes.
Wheel Diameter: 1372 mm. **Maximum Speed:** 15 mph.

x Dual (air/vacuum) braked. a Air-braked only.

13000–D 3000		Peak Rail	Derby 1952
13002–D 3002		Plym Valley Railway	Derby 1952
13014–D 3014	"SAMSON"	Dartmouth Steam Railway	Derby 1952
13018–D 3018–08011		Chinnor & Princes Risborough Railway	Derby 1953
13019–D 3019		Cambrian Railway Trust, Llynclys	Derby 1953
13022–D 3022–08015		Severn Valley Railway	Derby 1953
13023–D 3023–08016		Peak Rail	Derby 1953

13029–D 3029–08021		Tyseley Locomotive Works	Derby 1953
13030–D 3030–08022	"LION"	Cholsey & Wallingford Railway	Derby 1953
13044–D 3044–08032	"MENDIP"	Knights Rail Services, Eastleigh Works	Derby 1954
13059–D 3059–08046	"Brechin City"	Caledonian Railway	Derby 1954
13067–D 3067–08054		Embsay & Bolton Abbey Railway	Darlington 1953
13074–D 3074–08060	"UNICORN"	Cholsey & Wallingford Railway	Darlington 1953
13079–D 3079–08064		National Railway Museum, York (N)	Darlington 1954
13101–D 3101		Great Central Railway	Derby 1955
13167–D 3167–08102		Lincolnshire Wolds Railway	Derby 1955
13174–D 3174–08108	"Dover Castle"	Kent & East Sussex Railway	Derby 1955
13180–D 3180–08114		Nottingham Transport Heritage Centre	Derby 1955
13190–D 3190–08123		Cholsey & Wallingford Railway	Derby 1955
13201–D 3201–08133		Severn Valley Railway	Derby 1955
13232–D 3232–08164		East Lancashire Railway	Darlington 1956
13236–D 3236–08168		Bluebell Railway	Darlington 1956
13255–D 3255		*Location unknown*	Derby 1956
13261–D 3261		Swindon & Cricklade Railway	Derby 1956
13265–D 3265–08195	"MARK"	Llangollen Railway	Derby 1956
13290–D 3290–08220		Nottingham Transport Heritage Centre	Derby 1956
13308–D 3308–08238	"Charlie"	Dean Forest Railway	Darlington 1956
13336–D 3336–08266		Keighley & Worth Valley Railway	Darlington 1957
D 3358–08288	Unnumbered	Mid Hants Railway	Derby 1957
D 3429–08359		Chasewater Light Railway	Crewe 1958
D 3462–08377		West Somerset Railway	Darlington 1957
D 3551–08436 x		Swanage Railway	Derby 1958
D 3558–08443		Bo'ness & Kinneil Railway	Derby 1958
D 3559–08444		Bodmin & Wenford Railway	Derby 1958
D 3586–08471		Severn Valley Railway	Crewe 1958
D 3588–08473		Dean Forest Railway	Crewe 1958
D 3591–08476		Battlefield Railway	Crewe 1958
D 3594–08479		East Lancashire Railway	Horwich 1958
D 3605–08490		Strathspey Railway	Horwich 1958
D 3690–08528 x		Battlefield Railway	Horwich 1959
D 3723–08556		North Yorkshire Moors Railway	Darlington 1959
D 3757–08590 x	"RED LION"	Midland Railway-Butterley	Crewe 1959
D 3771–08604 x	604 "PHANTOM"	Didcot Railway Centre	Derby 1959
D 3795–08628 x		Bryn Engineering, Blackrod	Derby 1959
D 3798–08631 x	EAGLE	Mid Norfolk Railway	Derby 1959
D 3802–08635 x		Severn Valley Railway	Derby 1959
D 3861–08694 x		Great Central Railway	Horwich 1959
D 3867–08700 a		East Lancashire Railway	Horwich 1960
D 3902–08734 x		Dean Forest Railway	Crewe 1960
D 3935–08767 x		North Norfolk Railway	Horwich 1961
D 3937–08769	"Gladys"	Severn Valley Railway	Derby 1960
D 3940–08772 x	CAMULODUNUM	North Norfolk Railway	Derby 1960
D 3941–08773 x		Embsay & Bolton Abbey Railway	Derby 1960
D 3948–08780 x		Southall Depot, London	Derby 1960
D 3993–08825 a		Battlefield Railway	Derby 1960
D 3998–08830 a		Crewe Heritage Centre	Horwich 1960
D 4018–08850 x		North Yorkshire Moors Railway	Horwich 1961
D 4126–08896 x		Severn Valley Railway	Horwich 1962
D 4141–08911 x	"MATEY"	National Railway Museum, York	Horwich 1962
D 4145–08915 x		Stephenson Railway Museum	Horwich 1962
D 4157–08927 x		Gloucestershire Warwickshire Railway	Horwich 1962
D 4174–08944 x		East Lancashire Railway	Darlington 1962

▶ Class 07 0-6-0 D2994 (07010) is based at the Avon Valley Railway. On 10 July 2010 it is arriving at Bitton with a train from Avon Riverside. This was part of the former Midland Railway route to Bath.
Mark Few

▶ Class 08 0-6-0 08911 "MATEY" is used by the National Railway Museum for shunting. It is seen in the car park at York on 19 May 2009.
Paul Abell

▶ The Class 14s have proved to be much more successful in preservation and industrial use than with BR – all 56 class members having been withdrawn by BR within 5 years. Newly arrived at Murton on the Derwent Valley Light Railway on 21 May 2011 is maroon D9523.
Andrew Mason

CLASS 09 — 0-6-0

Built: 1959–62. Built at Darlington & Horwich. 26 built.
Engine: English Electric 6KT of 298 kW (400 hp) at 680 rpm.
Transmission: Electric: Two EE506 axle-hung traction motors.
Power at Rail: 201 kW (269 hp).
Maximum Tractive Effort: 111 kN (25 000 lbf).
Continuous Tractive Effort: 39 kN (8800 lbf) at 11.6 mph.
Weight: 50 tonnes. **Wheel Diameter:** 1372 mm.
Maximum Speed: 27 mph.

Dual (air/vacuum) braked.

D 3665–09001	Peak Rail	Darlington 1959
D 3668–09004	Swindon & Cricklade Railway	Darlington 1959
D 3721–09010	South Devon Railway	Darlington 1959
D 4103–09015	Rye Farm, Wishaw, Sutton Coldfield	Horwich 1961
D 4113–09025	East Kent Light Railway	Horwich 1962

CLASS 10 — 0-6-0

Built: 1955–62. Built at Darlington & Doncaster. 146 built.
Engine: Lister Blackstone ER6T of 261 kW (350 hp) at 750 rpm.
Transmission: Electric. Two GEC WT821 axle-hung traction motors.
Power at Rail: 198 kW (265 hp).
Maximum Tractive Effort: 156 kN (35 000 lbf). **Weight:** 47 tonnes.
Continuous Tractive Effort: 53.4 kN (12 000 lbf) at 8.2 mph.
Wheel Diameter: 1372 mm. **Maximum Speed:** 20 mph.

D 3452		Bodmin & Wenford Railway	Darlington 1957
D 3489	"COLONEL TOMLINE"	Spa Valley Railway	Darlington 1958
D 4067	"Margaret Ethel – Thomas Alfred Naylor"	Great Central Railway	Darlington 1961
D 4092		Barrow Hill Roundhouse	Darlington 1962

CLASS 14 — 0-6-0

Built: 1964–65 at Swindon. 56 built.
Engine: Paxman Ventura 6YJXL of 485 kW (650 hp) at 1500 rpm.
Transmission: Hydraulic. Voith L217u.
Maximum Tractive Effort: 135 kN (30 910 lbf).
Continuous Tractive Effort: 109 kN (26 690 lbf) at 5.6 mph.
Weight: 51 tonnes. **Wheel Diameter:** 1219 mm.
Maximum Speed: 40 mph. **Train Heating:** None.

x Dual (air/vacuum) braked

D 9500		Peak Rail	Swindon 1964
D 9502		Peak Rail	Swindon 1964
D 9513		Embsay & Bolton Abbey Railway	Swindon 1964
D 9516 x		Wensleydale Railway	Swindon 1964
D 9518	NCB 7	Nene Valley Railway	Swindon 1964
D 9520	45	Nene Valley Railway	Swindon 1964
D 9521		Dean Forest Railway	Swindon 1964
D 9523 x		Derwent Valley Light Railway	Swindon 1964
D 9524	14901	Gwili Railway	Swindon 1964
D 9525		Peak Rail	Swindon 1965
D 9526		West Somerset Railway	Swindon 1965
D 9531		East Lancashire Railway	Swindon 1965
D 9537		Rippingale Station, Lincolnshire	Swindon 1965
D 9539		Ribble Steam Railway	Swindon 1965
D 9551		Royal Deeside Railway	Swindon 1965
D 9553		Gloucestershire Warwickshire Railway	Swindon 1965
D 9555		Dean Forest Railway	Swindon 1965

D9524 has a Rolls Royce Type D8 cyl of 336kW (450 hp) engine which was fitted whilst in industrial use with BP. This engine was originally fitted to a Class 17.

A further two Class 14s are hired to industrial companies but may also return to the Nene Valley Railway or Kent & East Sussex Railway for maintenance – these are D9504 and D9529.

CLASS 15 — Bo-Bo

Built: 1957–60 by BTH/Clayton Equipment Company. 44 built.
Engine: Paxman 16YHXL of 597 kW (800 hp) at 1250 rpm.
Transmission: Electric. Four BTH 137AZ axle-hung traction motors.
Maximum Tractive Effort: 178 kN (40 000 lbf).
Continuous Tractive Effort: 88 kN (19 700 lbf) at 11.3 mph.
Weight: 69 tonnes. **Wheel Diameter**: 1003 mm.
Maximum Speed: 60 mph. **Train Heating**: None.

D 8233–ADB968001	East Lancashire Railway	BTH 1131/1960

CLASS 17 — Bo-Bo

Built: 1962–65 by Clayton Equipment Company. 117 built.
Engine: Two Paxman 67HXL of 336 kW (450 hp) at 1500 rpm.
Transmission: Electric. Four GEC WT421 axle-hung traction motors.
Power at Rail: 461 kW (618 hp).
Maximum Tractive Effort: 178 kN (40 000 lbf).
Continuous Tractive Effort 80 kN (18 000 lbf) at 12.8 mph.
Weight: 69 tonnes. **Wheel Diameter**: 1003 mm.
Maximum Speed: 60 mph. **Train Heating**: None.

D 8568	Chinnor & Princes Risborough Railway	CE 4365U/69 1964

CLASS 20 — Bo-Bo

Built: 1957–68 by English Electric at Vulcan Foundry, Newton-le-Willows or Robert Stephenson & Hawthorns, Darlington. 228 built.
Engine: English Electric 8SVT of 746 kW (1000 hp) at 850 rpm.
Transmission: Electric. Four EE 526/5D axle-hung traction motors.
Power at Rail: 574 kW (770 hp).
Maximum Tractive Effort: 187 kN (42 000 lbf).
Continuous Tractive Effort: 111 kN (25 000 lbf) at 11 mph.
Weight: 74 tonnes. **Wheel Diameter**: 1092 mm.
Maximum Speed: 75 mph. **Train Heating**: None.

Dual (air/vacuum) braked except D 8000.

D 8000–20050		National Railway Museum, York (N)	EE/VF 2347/D375 1957
D 8001–20001		Ecclesbourne Valley Railway	EE/VF 2348/D376 1957
D 8007–20007		Nottingham Transport Heritage Centre	EE/VF 2354/D382 1957
D 8020–20020		Bo'ness & Kinneil Railway	EE/RSH 2742/8052 1959
D 8031–20031		Keighley & Worth Valley Railway	EE/RSH 2753/8063 1960
D 8035–20035	2001	Colne Valley Railway	EE/VF 2757/D482 1959
D 8048–20048		Midland Railway-Butterley	EE/VF 2770/D495 1959
D 8059–20059		Severn Valley Railway	EE/RSH 2965/8217 1961
D 8063–20063	2002	Colne Valley Railway	EE/RSH 2969/8221 1961
D 8069–20069		Mid Norfolk Railway	EE/RSH 2975/8227 1961
D 8087–20087		East Lancashire Railway	EE/RSH 2993/8245 1961
D 8096–20096		Barrow Hill Roundhouse	EE/RSH 3003/8255 1961
D 8098–20098		Great Central Railway	EE/RSH 3004/8256 1961
D 8110–20110		South Devon Railway	EE/RSH 3016/8268 1962
D 8118–20118	Saltburn-by-the-Sea	North Norfolk Railway	EE/RSH 3024/8276 1962
D 8121–20121		Wensleydale Railway	EE/RSH 3027/8279 1962
D 8128–20228	2004	Barry Rail Centre	EE/VF 3599/D998 1966
D 8132–20132	Barrow Hill Depot	Barrow Hill Roundhouse	EE/VF 3603/D1002 1966
D 8137–20137	Murray B. Hofmeyr	Gloucestershire Warwickshire Railway	EE/VF 3608/D1007 1966
D 8142–20142		Barrow Hill Roundhouse	EE/VF 3613/D1013 1966

▲ Two views of the scenic Wensleydale Railway taken during a diesel gala on 9 April. 20020 and 20166 are passing Wensley with the 13.30 Leeming Bar–Redmire. **Andrew Mason**

▼ Visiting from Llangollen, D5310 (26010) is leaving Leyburn with the 13.20 Redmire–Leeming Bar. **Phil Chilton**

D 8154–20154	Nottingham Transport Heritage Centre	EE/VF 3625/D1024 1966
D 8166–20166	Wensleydale Railway	EE/VF 3637/D1036 1966
D 8169–20169	Stainmore Railway	EE/VF 3640/D1039 1966
D 8177–20177	Severn Valley Railway	EE/VF 3648/D1047 1966
D 8188–20188	Severn Valley Railway	EE/VF 3669/D1064 1967
D 8189–20189	London Underground, West Ruislip Depot	EE/VF 3670/D1065 1967
D 8305–20205	Midland Railway-Butterley	EE/VF 3686/D1081 1967
D 8314–20214	Lakeside & Haverthwaite Railway	EE/VF 3695/D1090 1967
D 8327–20227	London Underground, West Ruislip Depot	EE/VF 3685/D1080 1968

CLASS 24 Bo-Bo

Built: 1958–61 at Derby, Crewe & Darlington. 151 built.
Engine: Sulzer 6LDA28A of 870 kW (1160 hp) at 750 rpm.
Transmission: Electric. Four BTH 137BY axle-hung traction motors.
Power at Rail: 629 kW (843 hp).
Maximum Tractive Effort: 178 kN (40 000 lbf).
Continuous Tractive Effort: 95 kN (21 300 lbf) at 4.38 mph.
Weight: 78 (* 81) tonnes. **Wheel Diameter**: 1143 mm.
Maximum Speed: 75 mph. **Train Heating**: Steam.

D 5032–24032*		North Yorkshire Moors Railway	Crewe 1959
D 5054–24054–ADB 968008	"PHIL SOUTHERN"	East Lancashire Railway	Crewe 1959
D 5061–24061–RDB 968007-97201	EXPERIMENT	North Yorkshire Moors Railway	Crewe 1960
D 5081–24081		Gloucestershire Warwickshire Railway	Crewe 1960

CLASS 25 Bo-Bo

Built: 1961–67. Built at Darlington, Derby & Beyer Peacock, Manchester. 327 built.
Engine: Sulzer 6LDA28B of 930 kW (1250 hp) at 750 rpm.
Transmission: Electric. Four AEI 253AY axle-hung traction motors.
Power at Rail: 708 kW (949 hp).
Maximum Tractive Effort: 200 kN (45 000 lbf).
Continuous Tractive Effort: 93 kN (20 800 lbf) at 17.1 mph.
Weight: 72–76 tonnes. **Wheel Diameter**: 1143 mm.
Maximum Speed: 90 mph.

Class 25/1. Dual (air/vacuum) braked except D 5217. Train heating: Steam.

D 5185–25035	"CASTELL DINAS BRAN"	Great Central Railway	Darlington 1963
D 5207–25057		North Norfolk Railway	Derby 1963
D 5209–25059		Keighley & Worth Valley Railway	Derby 1963
D 5217–25067		Battlefield Railway	Derby 1963
D 5222–25072		Caledonian Railway	Derby 1963

Class 25/2. Dual (air/vacuum) braked except D 5233. Train heating: Steam: D 5233, D 7585/94. None: D7523/35/41.

D 5233–25083		Caledonian Railway	Derby 1963
D 7523–25173	"John F Kennedy"	West Somerset Railway	Derby 1965
D 7535–25185	"HERCULES"	Dartmouth Steam Railway	Derby 1965
D 7541–25191		South Devon Railway	Derby 1965
D 7585–25235		Bo'ness & Kinneil Railway	Darlington 1964
D 7594–25244		Kent & East Sussex Railway	Darlington 1964

Class 25/3. Dual (air/vacuum) braked. Train heating: None.

D 7612–25262–25901		South Devon Railway	Derby 1966
D 7615–25265		Great Central Railway	Derby 1966
D 7628–25278	"SYBILLA"	North Yorkshire Moors Railway	BP 8038/1965
D 7629–25279		Nottingham Transport Heritage Centre	BP 8039/1965
D 7633–25283–25904		Dean Forest Railway	BP 8043/1965
D 7659–25309–25909		West Coast Railway Co, Carnforth	BP 8069/1966
D 7663–25313		Wensleydale Railway	Derby 1966
D 7671–25321		Midland Railway-Butterley	Derby 1967
D 7672–25322–25912	TAMWORTH CASTLE	Churnet Valley Railway	Derby 1967

CLASS 26 Bo-Bo

Built: 1958–59 by the Birmingham Railway Carriage & Wagon Company. 47 built.
Engine: Sulzer 6LDA28B of 870 kW (1160 hp) at 750 rpm.
Transmission: Electric. Four Crompton-Parkinson C171A1 (§ C171D3) axle-hung traction motors.
Power at Rail: 671 kW (900 hp).
Maximum Tractive Effort: 187 kN (42 000 lbf).
Continuous Tractive Effort: 133 kN (30 000 lbf) at 14 mph.
Weight: 72–75 tonnes. **Wheel Diameter:** 1092 mm.
Maximum Speed: 75 mph.
Train Heating: Built with steam. Removed from D 5300/01/02/04 in 1967.
Dual (air/vacuum) braked.

Class 26/0.

D 5300–26007	Barrow Hill Roundhouse	BRCW DEL/45/1958
D 5301–26001	Caledonian Railway	BRCW DEL/46/1958
D 5302–26002	Strathspey Railway	BRCW DEL/47/1958
D 5304–26004	Bo'ness & Kinneil Railway	BRCW DEL/49/1958
D 5310–26010	Llangollen Railway	BRCW DEL/55/1959
D 5311–26011	Barrow Hill Roundhouse	BRCW DEL/56/1959
D 5314–26014	Caledonian Railway	BRCW DEL/59/1959

Class 26/1.§

D 5324–26024	Bo'ness & Kinneil Railway	BRCW DEL/69/1959
D 5325–26025	Strathspey Railway	BRCW DEL/70/1959
D 5335–26035	Caledonian Railway	BRCW DEL/80/1959
D 5338–26038	Pullman Rail, Cardiff Canton	BRCW DEL/83/1959
D 5340–26040	Barclay Brown Yard, Methil	BRCW DEL/85/1959
D 5343–26043	Gloucestershire Warwickshire Railway	BRCW DEL/88/1959

CLASS 27 Bo-Bo

Built: 1961–62 by the Birmingham Railway Carriage & Wagon Company, Birmingham. 69 built.
Engine: Sulzer 6LDA28B of 930 kW (1250 hp) at 750 rpm.
Transmission: Electric. Four GEC WT459 axle-hung traction motors.
Power at Rail: 696 kW (933 hp).
Maximum Tractive Effort: 178 kN (40 000 lbf).
Continuous Tractive Effort: 111 kN (25 000 lbf) at 14 mph.
Weight: 72–75 tonnes. **Wheel Diameter:** 1092 mm.
Maximum Speed: 90 mph.
Train Heating: Built with steam (except D 5370 – no provision). Replaced with electric on D 5386 and D 5410 but subsequently removed. Dual (air/vacuum) braked except D 5353.

D 5347–27001	Bo'ness & Kinneil Railway	BRCW DEL/190/1961
D 5351–27005	Bo'ness & Kinneil Railway	BRCW DEL/194/1961
D 5353–27007	Mid Hants Railway	BRCW DEL/196/1961
D 5370–27024–ADB 968028	Lakeside & Haverthwaite Railway	BRCW DEL/213/1962
D 5386–27103–27212–27066	Dean Forest Railway	BRCW DEL/229/1962
D 5394–27106–27050	Strathspey Railway	BRCW DEL/237/1962
D 5401–27112–27056	Great Central Railway	BRCW DEL/244/1962
D 5410–27123–27205–27059	Severn Valley Railway	BRCW DEL/253/1962

CLASS 28 METROVICK Co-Bo

Built: 1958–59 by Metropolitan Vickers, Manchester. 20 built.
Engine: Crossley HSTVee 8 of 896 kW (1200 hp) at 625 rpm.
Transmission: Electric. Five MV 137BZ axle-hung traction motors.
Power at Rail: 671 kW (900 hp).
Maximum Tractive Effort: 223 kN (50 000 lbf). **Weight:** 99 tonnes.
Continuous Tractive Effort: 111 kN (25 000 lbf) at 13.5 mph.
Wheel Diameter: 1003 mm. **Train Heating:** Steam.
Maximum Speed: 75 mph.

D 5705–S 15705–TDB 968006	East Lancashire Railway	MV 1958

▲ In a scene that could almost date from the early 1960s, D5185 (25035) "CASTELL DINAS BRAN" passes Woodthorpe with the 10.15 Loughborough–Leicester North on 12 September 2009, during the GCR's autumn diesel gala. **Robert Pritchard**

▲ D5862 (31327) shunts the stock from the Royal Scotsman luxury train at Boat of Garten on the Strathspey Railway on 29 May 2011 before hauling it up the line to Aviemore. **Jamie Squibbs**

▼ BR Blue 33109 passes Little Burrs with the 14.20 Rawtenstall–Bury on 7 March 2010. **Paul Senior**

CLASS 31 — A1A-A1A

Built: 1957–62 by Brush Electrical Engineering Company, Loughborough. 263 built.
Engine: Built with Mirrlees JVS12T of 930 kW (1250 hp). Re-engined 1964–69 with English Electric 12SVT of 1100 kW (1470 hp) at 850 rpm.
Transmission: Electric. Four Brush TM73-68 axle-hung traction motors.
Power at Rail: 872 kW (1170 hp).
Maximum Tractive Effort: 190 kN (42 800 lbf).
Continuous Tractive Effort: 99 kN (22 250 lbf) at 19.7 mph.
Weight: 110 tonnes. **Wheel Diameter:** 1092 mm.
Maximum Speed: 90 (§ 80) mph. D5518, D5522, D5526, and D5533 were originally 80 mph but have been regeared for 90 mph.
Train Heating: Built with steam heating. Electric heating fitted and steam heating removed on D5522, D5533, D5547, D5557, D5600, D5669, D5695, D5814 and D5830.
D5500, D5518, D5522, D5526, D5547 and D5562 were built without roof-mounted headcode boxes. They were subsequently fitted to D5518.

Class 31/0. Electromagnetic Control.

D5500–31018§		National Railway Museum, York (N)	BE 71/1957

Class 31/1. Electro-pneumatic Control. Dual (air/vacuum) braked.

D5518–31101		Battlefield Railway	BE 89/1958
D5522–31418		Midland Railway-Butterley	BE 121/1959
D5526–31108		Nene Valley Railway	BE 125/1959
D5533–31115–31466		Dean Forest Railway	BE 132/1959
D5537–31119		Embsay & Bolton Abbey Railway	BE 136/1959
D5546–31128	CHARYBDIS	North Yorkshire Moors Railway	BE 145/1959
D5547–31129–31461		Battlefield Railway	BE 146/1959
D5548–31130	Calder Hall Power Station	Battlefield Railway	BE 147/1959
D5557–31139–31438–31538		Mid Norfolk Railway	BE 156/1959
D5562–31144		Reliance Industrial Estate, Manchester	BE 161/1959
D5580–31162		Llangollen Railway	BE 180/1960
D5581–31163		Chinnor & Princes Risborough Railway	BE 181/1960
D5600–31179–31435		Embsay & Bolton Abbey Railway	BE 200/1960
D5627–31203	"Steve Organ G.M."	Pontypool & Blaenavon Railway	BE 227/1960
D5630–31206		Rushden Transport Museum	BE 230/1960
D5631–31207		North Norfolk Railway	BE 231/1960
D5634–31210		Dean Forest Railway	BE 234/1960
D5662–31235		Mid Norfolk Railway	BE 262/1960
D5669–31410	Granada Telethon	Stainmore Railway	BE 269/1960
D5683–31255		Colne Valley Railway	BE 284/1961
D5695–31265–31430–31530	Sister Dora	Mid Norfolk Railway	BE 296/1961
D5800–31270		Peak Rail	BE 301/1961
D5801–31271	"Stratford 1840–2001"	Midland Railway-Butterley	BE 302/1961
D5814–31414–31514		Ecclesbourne Valley Railway	BE 315/1961
D5821–31289	"PHOENIX"	Northampton & Lamport Railway	BE 322/1961
D5830–31297–31463–31563		Great Central Railway	BE 366/1962
D5862–31327	Phillips-Imperial	Strathspey Railway	BE 398/1962

31108 and 31162 are on loan from the Midland Railway-Butterley.

CLASS 33 — Bo-Bo

Built: 1961–62 by the Birmingham Railway Carriage & Wagon Company, Birmingham. 98 built.
Engine: Sulzer 8LDA28A of 1160 kW (1550 hp) at 750 rpm.
Transmission: Electric. Four Crompton-Parkinson C171C2 axle-hung traction motors.
Power at Rail: 906 kW (1215 hp).
Maximum Tractive Effort: 200 kN (45 000 lbf).
Continuous Tractive Effort: 116 kN (26 000 lbf) at 17.5 mph.
Weight: 78 tonnes. **Wheel Diameter:** 1092 mm.
Maximum Speed: 85 mph. **Train Heating:** Electric.

Dual (air/vacuum) braked.

§ fitted for push-pull operation (Class 33/1).
* built to former loading gauge of the Tonbridge–Battle line (Class 33/2).

D6501–33002	Sea King	South Devon Railway	BRCW DEL/93/1960
D6508–33008	Eastleigh	Battlefield Railway	BRCW DEL/100/1960
D6513–33102§		Churnet Valley Railway	BRCW DEL/105/1960
D6514–33103§	"SWORDFISH"	Swanage Railway	BRCW DEL/106/1960
D6515–33012		Swanage Railway	BRCW DEL/107/1960
D6521–33108§	"VAMPIRE"	Barrow Hill Roundhouse	BRCW DEL/113/1960
D6525–33109§	Captain Bill Smith RNR	East Lancashire Railway	BRCW DEL/117/1960
D6527–33110§		Bodmin & Wenford Railway	BRCW DEL/119/1960
D6528–33111§		Swanage Railway	BRCW DEL/120/1960
D6530–33018		Midland Railway-Butterley	BRCW DEL/122/1960
D6534–33019	Griffon	Battlefield Railway	BRCW DEL/126/1960
D6535–33116§	Hertfordshire Railtours	Great Central Railway (N)	BRCW DEL/127/1960
D6536–33117§		East Lancashire Railway	BRCW DEL/128/1960
D6539–33021		Churnet Valley Railway	BRCW DEL/131/1960
D6552–33034		Swanage Railway	BRCW DEL/144/1961
D6553–33035	Spitfire	Barrow Hill Roundhouse	BRCW DEL/145/1961
D6564–33046	Merlin	Midland Railway-Butterley	BRCW DEL/156/1961
D6566–33048		West Somerset Railway	BRCW DEL/170/1961
D6570–33052	Ashford	Kent & East Sussex Railway	BRCW DEL/174/1961
D6571–33053		Mid Hants Railway	BRCW DEL/175/1961
D6575–33057	Seagull	West Somerset Railway	BRCW DEL/179/1961
D6583–33063	"R.J. Mitchell DESIGNER OF THE SPITFIRE"	Spa Valley Railway	BRCW DEL/187/1962
D6585–33065	Sealion	Spa Valley Railway	BRCW DEL/189/1962
D6586–33201*		Midland Railway-Butterley	BRCW DEL/157/1962
D6587–33202*	"Dennis G. Robinson"	Mangapps Railway Museum	BRCW DEL/158/1962
D6593–33208*		Mid Hants Railway	BRCW DEL/164/1962

33202 also carried the names The Burma Star and METEOR.

CLASS 35 HYMEK B-B

Built: 1961–64 by Beyer Peacock, Manchester. 101 built.
Engine: Maybach MD 870 of 1269 kW (1700 hp) at 1500 rpm.
Transmission: Hydraulic. Mekydro K184U.
Maximum Tractive Effort: 207 kN (46 600 lbf).
Continuous Tractive Effort: 151 kN (33 950 lbf) at 12.5 mph.
Weight: 77 tonnes. **Wheel Diameter:** 1143 mm.
Maximum Speed: 90 mph. **Train Heating:** Steam.

D7017	West Somerset Railway	BP 7911/1962
D7018	West Somerset Railway	BP 7912/1962
D7029	Severn Valley Railway	BP 7923/1962
D7076	East Lancashire Railway	BP 7980/1963

CLASS 37 Co-Co

Built: 1960–66 by English Electric Company at Vulcan Foundry, Newton-le-Willows or Robert Stephenson & Hawthorns, Darlington. 309 built.
Engine: English Electric 12CSVT of 1300 kW (1750 hp) at 850 rpm (y re-engined with Ruston RK270T of 1340 kW (1800 hp at 900 rpm) (z re-engined with Mirrlees MB275T of 1340 kW (1800 hp) at 1000 rpm).
Transmission: Electric. Six English Electric 538/A.
Power at Rail: 932 kW (1250 hp).
Maximum Tractive Effort: 245 kN (55 000 lbf).
Continuous Tractive Effort: 156 kN (35 000 lbf) at 13.6 mph.
Weight: 103–108 tonnes. (y z 120 tonnes) **Wheel Diameter:** 1092 mm.
Maximum Speed: 80 mph.
Train Heating: Steam (§ possibly built without heating, * built without heating, but steam later fitted to D6963 and D6964). Provision of steam heat was removed from D6776, D6823, D6836, D6842, D6846, D6850, D6852, D6859, D6869, D6906, D6907, D6915, D6916, D6919, D6927, D6940, D6948, D6950, D6954, D6955 soon after delivery and may never have been used.

▲ D6515 (33012) leaves Harmans Cross with the 15.05 to Norden during the Swanage Railway diesel gala on 6 May 2011. The loco had just been returned to use following a major overhaul.
Tom McAtee

▼ One of the two "Hymeks" based at the West Somerset Railway, D7017, slows for the token exhange at Williton with the 18.00 Bishops Lydeard–Minehead on 13 June 2010. **Lindsay Atkinson**

▲ In original BR Green, D6737 (37037) crosses Mytholmes Viaduct on the Keighley & Worth Valley Railway with a Keighley–Oxenhope train on 8 June 2008. **Andrew Mason**

▼ BR Railfreight-liveried 37518 leaves Heywood with a Rawtenstall train on 6 March 2011. **Andrew Mason**

Dual (air/vacuum) braked.

D 6605–37305–37407	Loch Long	Churnet Valley Railway	EE/VF 3565/D994/1965
D 6607–37307–37403	Isle of Mull	Bo'ness & Kinneil Railway	EE/VF 3567/D996/1965
D 6608–37308–37274–37308		Pullman Rail, Cardiff Canton	EE/VF 3568/D997/1966
D 6700–37119–37350	NATIONAL RAILWAY MUSEUM	National Railway Museum, York (N)	EE/VF 2863/D579/1960
D 6703–37003		Mid Norfolk Railway	EE/VF 2866/D582/1960
D 6709–37009–37340		Nottingham Transport Heritage Centre	EE/VF 2872/D588/1961
D 6723–37023	Stratford TMD	Allely's, The Slough, Studley	EE/VF 2886 D602/1961
D 6725–37025	Inverness TMD	Bo'ness & Kinneil Railway	EE/VF 2888/D604/1961
D 6729–37029		Epping–Ongar Railway	EE/VF 2892/D608/1961
D 6732–37032–37353		North Norfolk Railway	EE/VF 2895/D611/1962
D 6737–37037–37321	Gartcosh	North Norfolk Railway	EE/VF 2900/D616/1962
D 6742–37042		Eden Valley Railway	EE/VF 3034/D696/1962
D 6757–37057	Viking	Barrow Hill Roundhouse	EE/VF 3049/D711/1962
D 6775–37075		Churnet Valley Railway	EE/RSH 3067/8321/1962
D 6776–37076–37518		East Lancashire Railway	EE/RSH 3068/8322/1962
D 6797–37097		Caledonian Railway	EE/VF 3226/D751/1962
D 6799–37099–37324	Clydebridge	Gloucestershire Warwickshire Railway	EE VF 3228/D753/1962
D 6808–37108–37325	Lanarkshire Steel	Crewe Heritage Centre	EE/VF 3237/D762/1963
D 6809–37109		East Lancashire Railway	EE VF 3238/D763/1963
D 6816–37116	Sister Dora	Chinnor & Princes Risborough Railway	EE/VF 3245/D770/1963
D 6823–37123–37679§		Northampton & Lamport Railway	EE/RSH 3268/8383/1963
D 6836–37136–37905y§	Vulcan Enterprise	Mid Hants Railway	EE/VF 3281/D810/1963
D 6842–37142§		Bodmin & Wenford Railway	EE/VF 3317/D816/1963
D 6846–37146§		Stainmore Railway	EE/VF 3321/D821/1963
D 6850–37150–37901z§	Mirrlees Pioneer	East Lancashire Railway	EE/VF 3325/D824/1963
D 6852–37152–37310§	British Steel Ravenscraig	Peak Rail	EE/VF 3327/D826/1963
D 6859–37159–37372		Barrow Hill Roundhouse	EE/RSH 3337/8390/1963
D 6869–37169–37674§	Saint Blaise Church 1445–1995	Stainmore Railway	EE/VF 3347/D833/1963
D 6875–37175		Bo'ness & Kinneil Railway	EE/VF 3353/D839/1963
D 6888–37188	Jimmy Shand	Peak Rail	EE/RSH 3366/8409/1963
D 6890–37190–37314	Dalzell	Midland Railway-Butterley	EE/RSH 3368/8411/1964
D 6906–37206–37906y§		Severn Valley Railway	EE/VF 3384/D850/1963
D 6907–37207§	William Cookworthy	Plym Valley Railway	EE/VF 3385/D851/1963
D 6915–37215§		Gloucestershire Warwickshire Railway	EE/VF 3393/D859/1964
D 6916–37216§	Great Eastern	Pontypool & Blaenavon Railway	EE/VF 3394/D860 1964
D 6919–37219§	"Shirley Ann Smith"	Mid Norfolk Railway	EE/VF 3405/D863/1964
D 6927–37227§		Battlefield Railway	EE/VF 3413/D871/1964
D 6940–37240*		Llangollen Railway	EE/VF 3497/D928/1964
D 6948–37248	Loch Arkaig	Gloucestershire Warwickshire Railway	EE/VF 3505/D936/1964
D 6950–37250		Eden Valley Railway	EE/VF 3507/D938/1964
D 6954–37254*	"Driver Robin Prince M.B.E."	Spa Valley Railway	EE/VF 3511/D942/1965
D 6955–37255*		Great Central Railway	EE/VF 3512/D943/1965
D 6963–37263*		Dean Forest Railway	EE/VF 3523/D952/1965
D 6964–37264*		North Yorkshire Moors Railway	EE/VF 3524/D953/1965
D 6967–37267–37421	Strombidae	Pontypool & Blaenavon Railway	EE/VF 3527/D956/1965
D 6971–37271–37418	Pectinidae	East Lancashire Railway	EE/VF 3531/D960/1965
D 6975–37275*	Stainless Pioneer	Barrow Hill Roundhouse	EE/VF 3535/D964/1965
D 6976–37276–37413	Loch Eil Outward Bound	Bo'ness & Kinneil Railway	EE/VF 3536/D965/1965
D 6979–37279–37424	Isle of Mull	Churnet Valley Railway	EE/VF 3539/D968/1965
D 6994–37294		Embsay & Bolton Abbey Railway	EE/VF 3554/D983/1965

37372 is undergoing conversion to a replica Class 23 "Baby Deltic", none of which were preserved.

D 6703 also carried the name First East Anglian Regiment for a short time in 1963.
37248 also carried the name Midland Railway Centre.
37275 also carried the name Oor Wullie.

37403 also carried the names Ben Cruachan and Glendarroch.
37407 also carried the name Blackpool Tower.
37413 also carried the name The Scottish Railway Preservation Society.
37418 also carried the names An Comunn Gaidhealach and East Lancashire Railway.
37421 also carried the names The Kingsman and Star of the East.
37424 also carried the name Glendarroch.

37037 is on loan from the South Devon Railway
37905 is on loan from the Battlefield Railway.

CLASS 40 — 1Co-Co1

Built: 1958–62 by English Electric at Vulcan Foundry, Newton-le-Willows & Robert Stephenson & Hawthorns, Darlington. 200 built.
Engine: English Electric 16SVT MkII of 1480 kW (2000 hp) at 850 rpm.
Transmission: Electric. Six EE 526/5D axle-hung traction motors.
Power at Rail: 1156 kW (1550 hp).
Maximum Tractive Effort: 231 kN (52 000 lbf).
Continuous Tractive Effort: 137 kN (30 900 lbf) at 18.8 mph.
Weight: 132 tonnes.
Wheel Diameters: 914/1143 mm.
Maximum Speed: 90 mph.
Train Heating: Steam.

Dual (air/vacuum) braked except D306.

D200–40122		National Railway Museum, York (N)	EE/VF 2367/D395 1958
D212–40012–97407	AUREOL	Midland Railway-Butterley	EE/VF 2667/D429 1959
D213–40013	ANDANIA	Barrow Hill Roundhouse	EE/VF 2668/D430 1959
D306–40106	"ATLANTIC CONVEYOR"	Boden Rail Engineering, Washwood Heath	EE/RSH 2726/8136 1960
D318–40118–97408		Tyseley Locomotive Works	EE/RSH 2853/8148 1961
D335–40135–97406		East Lancashire Railway	EE/VF 3081/D631 1961
D345–40145	"East Lancashire Railway"	East Lancashire Railway	EE/VF 3091/D641 1961

CLASS 42 — WARSHIP — B-B

Built: 1958–61 at Swindon. 38 Built.
Engines: Two Maybach MD650 of 821 kW (1100 hp) at 1530 rpm.
Transmission: Hydraulic. Mekydro K 104U.
Maximum Tractive Effort: 223 kN (52 400 lbf).
Continuous Tractive Effort: 209 kN (46 900 lbf) at 11.5 mph.
Weight: 80 tonnes.
Wheel Diameter: 1033 mm.
Maximum Speed: 90 mph.
Train Heating: Steam.

D821	GREYHOUND	Severn Valley Railway	Swindon 1960
D832	ONSLAUGHT	West Somerset Railway	Swindon 1961

D832 is on loan from the East Lancashire Railway.

CLASS 44 — PEAK — 1Co-Co1

Built: 1959–60 at Derby. 10 Built.
Engine: Sulzer 12LDA28A of 1720 kW (2300 hp) at 750 rpm.
Transmission: Electric. Six Crompton Parkinson C171B1 axle-hung traction motors.
Power at Rail: 1342 kW (1800 hp).
Maximum Tractive Effort: 222 kN (50 000 lbf).
Continuous Tractive Effort: 129 kN (29 100 lbf) at 23.2 mph.
Weight: 135 tonnes.
Wheel Diameters: 914/1143 mm.
Maximum Speed: 90 mph.
Train Heating: Built with steam but facility removed in 1962.

D4–44004	GREAT GABLE	Midland Railway-Butterley	Derby 1959
D8–44008	PENYGHENT	Peak Rail	Derby 1959

▲ 40145 "East Lancashire Railway" leads 37418 "Pectinidae" through Burrs with the 14.00 Rawtenstall–Heywood on 9 January 2010. **Phil Chilton**

▼ The two preserved "Warships", D832 "ONSLAUGHT" and D821 "GREYHOUND", have just left Blue Anchor with the 13.00 Bishops Lydeard–Minehead on 11 June 2010. **Lindsay Atkinson**

▲ 45041 "ROYAL TANK REGIMENT" arrives at Swanwick with the 10.48 Hammersmith–Riddings on 2 October 2010. **Tom McAtee**

▼ D1501 (47402) leaves Irwell Vale with the 10.40 Rawtenstall–Heywood on 9 January 2010. **Robert Pritchard**

CLASS 45 1Co-Co1

Built: 1960–63 at Crewe and Derby. 127 built.
Engine: Sulzer 12LDA28B of 1860 kW (2500 hp) at 750 rpm.
Transmission: Electric. Six Crompton Parkinson C172A1 axle-hung traction motors.
Power at Rail: 1491 kW (2000 hp).
Maximum Tractive Effort: 245 kN (55 000 lbf).
Continuous Tractive Effort: 133 kN (30 000 lbf) at 25 mph.
Weight: 135 tonnes. (D 14, D 53 and D 100 138 tonnes).
Wheel Diameters: 914/1143 mm. **Maximum Speed:** 90 mph.
Train Heating: Built with steam but subsequently replaced with electric except D 14, D 53 and D 100.

Dual (air/vacuum) braked.

D 14–45015		Battlefield Railway	Derby 1960
D 22–45132		Mid Hants Railway	Derby 1961
D 40–45133		Midland Railway-Butterley	Derby 1961
D 53–45041	ROYAL TANK REGIMENT	Midland Railway-Butterley	Crewe 1962
D 61–45112	THE ROYAL ARMY ORDNANCE CORPS	Barrow Hill Roundhouse	Crewe 1962
D 67–45118	THE ROYAL ARTILLERYMAN	RVEL, Derby	Crewe 1962
D 86–45105		Barrow Hill Roundhouse	Crewe 1961
D 99–45135	3rd CARABINIER	East Lancashire Railway	Crewe 1961
D 100–45060	SHERWOOD FORESTER	Barrow Hill Roundhouse	Crewe 1961
D 120–45108		Midland Railway-Butterley	Crewe 1961
D 123–45125	"LEICESTERSHIRE AND DERBYSHIRE YEOMANRY"	Great Central Railway	Crewe 1961
D 135–45149		Gloucestershire Warwickshire Railway	Crewe 1961

CLASS 46 1Co-Co1

Built: 1961–63 at Derby. 56 built.
Engine: Sulzer 12LDA28B of 1860 kW (2500 h.p.) at 750 rpm.
Transmission: Electric. Six Brush TM73-68 MkIII axle-hung traction motors.
Power at Rail: 1460 kW (1960 hp).
Maximum Tractive Effort: 245 kN (55 000 lbf).
Continuous Tractive Effort: 141 kN (31 600 lbf) at 22.3 mph.
Weight: 141 tonnes. **Wheel Diameters:** 914/1143 mm.
Maximum Speed: 90 mph.
Train Heating: Steam.

Dual (air/vacuum) braked.

D 147–46010		Nottingham Transport Heritage Centre	Derby 1961
D 172–46035–97403	Ixion	Crewe Heritage Centre	Derby 1962
D 182–46045–97404		Midland Railway-Butterley	Derby 1962

CLASS 47 Co-Co

Built: 1963–67 at Crewe and Brush Electrical Engineering Company, Loughborough. 512 built.
Engine: Sulzer 12LDA28C of 1920 kW (2580 hp) at 750 rpm.
Transmission: Electric. Six Brush TG 160-60 axle-hung traction motors.
Power at Rail: 1550 kW (2080 hp).
Maximum Tractive Effort: 245 kN (55 000 lbf). (§ 267 kN (60 000 lbf)).
Continuous Tractive Effort: 133 kN (33 000 lbf) at 26 mph.
Weight: 119–121 tonnes. **Wheel Diameters:** 1143 mm.
Maximum Speed: 95 mph (D 1932, D 1945 and D 1960 100 mph).
Train Heating: Built with steam except D 1787, D 1886, D 1894 and D 1895 no heat and D 1500, D 1501 & D 1516 steam/electric. Electric heating fitted & steam heating removed on D 1107, D 1566, D 1606, D 1619, D 1643, D 1656, D 1661, D 1662, D 1713, D 1723, D 1754, D 1755, D 1762, D 1778, D 1909, D 1921, D 1927, D 1932, D 1933, D 1945, D 1946, D 1960 and D 1970.

D 1705 was built with a Sulzer 12LVA24 engine but replaced with a 12LDA28C in 1972.

Dual (air/vacuum) braked.

D1107–47524§	Res Gestae	Churnet Valley Railway	Crewe 1966
D1500–47401	North Eastern	Midland Railway-Butterley	BE 342/1962
D1501–47402	Gateshead	East Lancashire Railway	BE 343/1962
D1516–47417		Midland Railway-Butterley	BE 358/1963
D1524–47004§	Old Oak Common Traction & Rolling Stock Depot	Embsay & Bolton Abbey Railway	BE 419/1963
D1566–47449§	"ORION"	Llangollen Railway	Crewe 1964
D1606–47029–47635	The Lass O' Ballochmyle	Peak Rail	Crewe 1964
D1619–47038–47564–47761	COLOSSUS	Midland Railway-Butterley	Crewe 1964
D1643–47059–47631–47765§	Ressaldar	Nottingham Transport Heritage Centre	Crewe 1965
D1656–47072–47609–47798	Prince William	National Railway Museum, York (N)	Crewe 1965
D1661–47077–47613–47840	NORTH STAR	West Somerset Railway	Crewe 1965
D1662–47484§	ISAMBARD KINGDOM BRUNEL	Rye Farm, Wishaw, Sutton Coldfield	Crewe 1965
D1693–47105§		Gloucestershire Warwickshire Railway	BE 455/1963
D1705–47117§	"SPARROWHAWK"	Great Central Railway	BE 467/1965
D1713–47488	Rail Riders	Barrow Hill Roundhouse	BE 475/1964
D1723–47540–47975–47540§	The Institution of Civil Engineers	Wensleydale Railway	BE 494/1964
D1754–47160–47605–47746	The Bobby	St Modwen Properties, Long Marston	BE 482/1964
D1755–47541–47773	The Queen Mother	Tyseley Locomotive Works	BE 483/1964
D1762–47167–47580–47732	County of Essex	Mid Norfolk Railway	BE 524/1964
D1778–47183–47579–47793	Christopher Wren	Mangapps Railway Museum	BE 540/1964
D1787–47306	The Sapper	Bodmin & Wenford Railway	BE 549/1964
D1842–47192§		Crewe Heritage Centre	Crewe 1965
D1855–47205§		Northampton & Lamport Railway	Crewe 1965
D1886–47367§		North Norfolk Railway	BE 648/1965
D1894–47375§	Tinsley Traction Depot	Barrow Hill Roundhouse	BE 657/1965
D1895–47376§	Freightliner 1995	Gloucestershire Warwickshire Railway	BE 657/1965
D1909–47232–47665–47785	Fiona Castle	Stainmore Railway	BE 671/1965
D1921–47244–47640§	University of Strathclyde	Battlefield Railway	BE 683/1966
D1927–47250–47600–47744	Royal Mail Cheltenham	Barrow Hill Roundhouse	BE 689/1966
D1932–47493–47701	Saint Andrew	Dartmoor Railway	BE 694/1966
D1933–47255–47596§	Aldeburgh Festival	Mid Norfolk Railway	BE 695/1966
D1945–47502–47715	Haymarket	Wensleydale Railway	BE 707/1966
D1946–47503–47771§	Heaton Traincare Depot	Colne Valley Railway	BE 708/1966
D1960–47514–47703	Saint Mungo	Wensleydale Railway	BE 622/1967
D1970–47269–47643§		Bo'ness & Kinneil Railway	Crewe 1965
D1994–47292		Nottingham Transport Heritage Centre	Crewe 1966

47401 also carried the name Star of the East.
47488 also carried the name DAVIES THE OCEAN.
47490 also carried the name Bristol Bath Road.
47600/744 also carried the names Dewi Saint/Saint David, Saint Edwin and The Cornish Experience.
47635 also carried the name Jimmy Milne.
47701 also carried the names Waverley and Old Oak Common Traction & Rolling Stock Depot.
47703 also carried the names The Queen Mother, LEWIS CARROLL and HERMES.
47715 now carries the name POSEIDON.
47732 also carried the name Restormel.
47771 also carried the name The Geordie.
47773 also carried the name Reservist.
47785 also carried the name The Statesman.
47793 also carried the names James Nightall G.C. and Saint Augustine.
47798 also carried the name FIRE FLY.

47798 was also numbered 47834 for a time and 47785 was numbered 47820 for a time.

▲ 50007 "SIR EDWARD ELGAR" passes Woodthorpe with the 14.25 Loughborough–Leicester North on 12 September 2009. **Lindsay Atkinson**

▼ Part of the National Collection, BR Blue D1023 "WESTERN FUSILIER" arrives at Goathland with the 12.50 Grosmont–Pickering on 19 September 2009. **Andrew Mason**

CLASS 50 Co-Co

Built: 1967–68 by English Electric at Vulcan Foundry, Newton-le-Willows. 50 built.
Engine: English Electric 16CVST of 2010 kW (2700 hp) at 850 rpm.
Transmission: Electric. Six EE 538/5A axle-hung traction motors.
Power at Rail: 1540 kW (2070 hp).
Maximum Tractive Effort: 216 kN (48 500 lbf).
Continuous Tractive Effort: 147 kN (33 000 lbf) at 18.8 mph.
Weight: 117 tonnes. **Wheel Diameter:** 1092 mm.
Maximum Speed: 100 mph.
Train Heating: Electric.

Dual (air/vacuum) braked.

D400–50050	Fearless	Yeovil Railway Centre	EE/VF 3770/D1141 1967
D402–50002	Superb	South Devon Railway	EE/VF 3772/D1143 1967
D407–50007	SIR EDWARD ELGAR	Midland Railway-Butterley	EE/VF 3777/D1148 1968
D408–50008	Thunderer	Boden Rail Engineering, Washwood Heath	EE/VF 3778/D1149 1968
D415–50015	Valiant	East Lancashire Railway	EE/VF 3785/D1156 1968
D417–50017	Royal Oak	Plym Valley Railway	EE/VF 3787/D1158 1968
D419–50019	Ramillies	Mid Norfolk Railway	EE/VF 3789/D1160 1968
D421–50021	Rodney	Tyseley Locomotive Works	EE/VF 3791/D1162 1968
D426–50026	Indomitable	Severn Valley Railway	EE/VF 3796/D1167 1968
D427–50027	Lion	North Yorkshire Moors Railway	EE/VF 3797/D1168 1968
D429–50029	Renown	Peak Rail	EE/VF 3799/D1170 1968
D430–50030	Repulse	Peak Rail	EE/VF 3800/D1171 1968
D431–50031	Hood	Severn Valley Railway	EE/VF 3801/D1172 1968
D433–50033	Glorious	Tyseley Locomotive Works	EE/VF 3803/D1174 1968
D435–50035–50135	Ark Royal	Severn Valley Railway	EE/VF 3805/D1176 1968
D442–50042	Triumph	Bodmin & Wenford Railway	EE/VF 3812/D1183 1968
D444–50044	Exeter	Pullman Rail, Cardiff Canton	EE/VF 3814/D1185 1968
D449–50049–50149	Defiance	Pullman Rail, Cardiff Canton	EE/VF 3819/D1190 1968

50007 was previously named Hercules.

CLASS 52 WESTERN C-C

Built: 1961–64 at Crewe and Swindon. 74 built.
Engines: Two Maybach MD655 of 1007 kW (1350 hp) at 1500 rpm.
Transmission: Hydraulic. Voith L630rV.
Maximum Tractive Effort: 297.3 kN (66 770 lbf).
Continuous Tractive Effort: 201.2 kN (45 200 lbf) at 14.5 mph.
Weight: 111 tonnes. **Wheel Diameter:** 1092 mm.
Maximum Speed: 90 mph. **Train Heating:** Steam.

Dual (air/vacuum) braked.

D1010	WESTERN CAMPAIGNER	West Somerset Railway	Swindon 1962
D1013	WESTERN RANGER	Severn Valley Railway	Swindon 1962
D1015	WESTERN CHAMPION	Severn Valley Railway	Swindon 1963
D1023	WESTERN FUSILIER	National Railway Museum, York (N)	Swindon 1963
D1041	WESTERN PRINCE	East Lancashire Railway	Crewe 1962
D1048	WESTERN LADY	Midland Railway-Butterley	Crewe 1962
D1062	WESTERN COURIER	Severn Valley Railway	Crewe 1963

CLASS 55 DELTIC Co-Co

Built: 1961–62 by English Electric at Vulcan Foundry, Newton-le-Willows. 22 built.
Engine: Two Napier Deltic T18-25 of 1230 kW (1650 hp) at 1500 rpm.
Transmission: Electric. Six EE 538 axle-hung traction motors.
Power at Rail: 1969 kW (2640 hp).
Maximum Tractive Effort: 222 kN (50 000 lbf).
Continuous Tractive Effort: 136 kN (30 500 lbf) at 32.5 mph.
Weight: 105 tonnes. **Wheel Diameter:** 1092 mm.
Maximum Speed: 100 mph.
Train Heating: Built with Steam, electric subsequently fitted.

Dual (air/vacuum) braked.

D 9000–55022	ROYAL SCOTS GREY	East Lancashire Railway	EE/VF 2905/D557 1961
D 9002–55002	THE KING'S OWN YORKSHIRE LIGHT INFANTRY	National Railway Museum, York (N)	EE/VF 2907/D559 1961
D 9009–55009	ALYCIDON	Barrow Hill Roundhouse	EE/VF 2914/D566 1961
D 9015–55015	TULYAR	Barrow Hill Roundhouse	EE/VF 2920/D572 1961
D 9016–55016	GORDON HIGHLANDER	East Lancashire Railway	EE/VF 2921/D573 1961
D 9019–55019	ROYAL HIGHLAND FUSILIER	Barrow Hill Roundhouse	EE/VF 2924/D576 1961

CLASS 56 Co-Co

Built: 1976–84 by Electroputere, Craiova, Romania and BREL Doncaster & Crewe. 135 built.
Engine: Ruston-Paxman 16RK3CT of 2460 kW (3250 hp) at 900 rpm.
Transmission: Electric. Six Brush TM73-62 axle-hung traction motors.
Power at Rail: 1790 kW (2400 hp).
Maximum Tractive Effort: 275 kN (61 800 lbf).
Continuous Tractive Effort: 240 kN (53 950 lbf) at 32.5 mph.
Weight: 125 tonnes. **Wheel Diameter:** 1143 mm.
Maximum Speed: 80 mph.
Train Heating: None.

Air braked.

56045–56301	British Steel Shelton	Barrow Hill Roundhouse	Doncaster 1978
56086	The Magistrates' Association	Battlefield Railway	Doncaster 1980
56097		Nottingham Transport Heritage Centre	Doncaster 1981
56098		Battlefield Railway	Doncaster 1981
56101	Mutual Improvement	Mid Norfolk Railway	Doncaster 1981
56124–56302	Wilson Walshe	Barrow Hill Roundhouse	Crewe 1983

56124 also carried the name Blue Circle Cement.

CLASS 58 Co-Co

Built: 1983–87 by BREL Doncaster. 50 built.
Engine: Ruston-Paxman 16RK3ACT of 2460 kW (3300 hp) at 1000 rpm
Transmission: Electric. Brush TM73-62 traction motors.
Power at Rail: 1780 kW (2387 hp).
Maximum Tractive Effort: 275 kN (61 800 lbf).
Continuous Tractive Effort: 240 kN (53 950 lbf) at 32.5 mph.
Weight: 130 tonnes. **Wheel Diameter:** 1120 mm.
Maximum Speed: 80 mph. **Train Heating:** None.

Air braked.

58016	Barrow Hill Roundhouse	Doncaster 1984

▲ BR Green D9016 "GORDON HIGHLANDER" has just crossed the Metrolink line at Bury with the 10.40 Rawtenstall–Heywood on 22 April 2011. **Tom McAtee**

▼ Newly preserved 56301, in Fastline livery, leaves Corfe Castle with the 13.10 Norden–Swanage on 8 May 2011. **Mark Few**

CLASS 98/1 0-6-0

Built: 1987 by Brecon Mountain Railway. 1 built for use on Aberystwyth–Devil's Bridge line.
Engine: Caterpillar 3304T of 105 kW (140 hp).
Transmission: Hydraulic. Twin Disc torque converter.
Gauge: 1′ 11½″. **Weight:** 12.75 tonnes.
Maximum Speed: 15 mph. **Wheel Diameter:** 610 mm.

10 Vale of Rheidol Railway BMR 1987

UNCLASSIFIED HUDSWELL-CLARKE 0-6-0

Built: 1955–61 by Hudswell-Clarke & Company, Leeds. 20 built.
Engine: Gardner 8L3 of 152 kW (204 hp) at 1200 rpm.
Transmission: Mechanical. SSS powerflow double synchro.
Maximum Tractive Effort: 85.7 kN (19 245 lbf).
Continuous Tractive Effort: 76 kN (17 069 lbf) at 3.72 mph.
Weight: 34 tonnes. **Wheel Diameter:** 1067 mm.
Maximum Speed: 25 mph.

D 2511 Keighley & Worth Valley Railway HC D 1202/1961

UNCLASSIFIED NORTH BRITISH 0-4-0

Built: 1957–61 by North British Locomotive Company, Glasgow. 73 built.
Engine: MAN W6V 17.5/22A of 168 kW (225 hp) at 1100 rpm.
Transmission: Mechanical. Voith L33YU.
Maximum Tractive Effort: 89.4 kN (20 080 lbf).
Continuous Tractive Effort: 53.4 kN (12 000 lbf) at 4 mph.
Weight: 28 tonnes. **Wheel Diameter:** 1067 mm.
Maximum Speed: 15 mph.

D 2767 Bo'ness & Kinneil Railway NBL 28020/1960
D 2774 Strathspey Railway NBL 28027/1960

2.4. EXPERIMENTAL DIESEL LOCOMOTIVES

PROTOTYPE DELTIC Co-Co

Built: 1955 by English Electric. Used by BR 1959–61.
Engine: Two Napier Deltic T18-25 of 1230 kW (1650 hp) at 1500 rpm.
Transmission: Electric. Six EE 526A axle-hung traction motors.
Power at Rail: 1976 kW (2650 hp).
Maximum Tractive Effort: 267 kN (60 000 lbf).
Continuous Tractive Effort: 104 kN (23 400 lbf) at 43.5 mph.
Weight: 107.7 tonnes. **Wheel Diameter:** 1092 mm.
Maximum Speed: 105 mph. **Train Heating:** Steam.

DELTIC National Railway Museum, Shildon (N) EE 2003/1955

PROTOTYPE ENGLISH ELECTRIC SHUNTER 0-6-0

Built: 1957 by English Electric at Vulcan Foundry, Newton-le-Willows. Used by BR 1957–60.
Engine: English Electric 6RKT of 373 kW (500 hp) at 750 rpm.
Transmission: Electric.
Power at Rail:
Maximum Tractive Effort: 147 kN (33 000 lbf).
Continuous Tractive Effort: (lbf) at mph.
Weight: 48 tonnes. **Wheel Diameter:** 1219 mm.
Maximum Speed: 35 mph.

D 226–D 0226 "VULCAN" Keighley & Worth Valley Railway EE/VF 2345/D226 1956

PROTOTYPE NORTH BRITISH SHUNTER 0-4-0

Built: 1954 by North British Locomotive Company, Glasgow. Used by BR Western Region (27414) and BR London Midland & Southern regions (27415). Subsequently sold for industrial use.
Engine: Paxman 6 VRPHXL of 160 kW (225 hp) at 1250 rpm.
Transmission: Hydraulic. Voith L24V. **Weight:** ?? tonnes.
Maximum Tractive Effort: 112 kN (22 850 lbf). **Maximum Speed:** 12 mph.
Wheel Diameter: 1016 mm.

No train brakes.

BR	*Present*		
–	TOM	Telford Steam Railway	NBL 27414/1954
–	TIGER	Bo'ness & Kinneil Railway	NBL 27415/1954

PROTOTYPE NORTH BRITISH SHUNTER 0-4-0

Built: 1958 by North British Locomotive Company, Glasgow. Used by BR Western Region at Old Oak Common in 1958. Subsequently sold for industrial use.
Engine: MAN W6V 17.5/22 of 168 kW (225 hp).
Transmission: Hydraulic. Voith L24V. **Weight:** ?? tonnes.
Maximum Tractive Effort: 112 kN (22 850 lbf). **Maximum Speed:** 12 mph.
Wheel Diameter: 940 mm.

No train brakes.

BR	*Present*		
–	D1	The Pallot Heritage Steam Museum, Jersey	NBL 27734/1958

2.5. CIVIL ENGINEERS DIESEL LOCOMOTIVES

CLASS 97/6 0-6-0

Built: 1952–59 by Ruston & Hornsby at Lincoln for BR Western Region Civil Engineers. 5 built.
Engine: Ruston 6VPH of 123 kW (165 hp).
Transmission: Electric. One British Thomson Houston RTA5041 traction motor.
Maximum Tractive Effort: 75 kN (17 000 lbf). **Weight:** 31 tonnes.
Maximum Speed: 20 mph. **Wheel Diameter:** 978 mm.

PWM 650–97650	Lincolnshire Wolds Railway	RH 312990/1952
PWM 651–97651	Strathspey Railway	RH 431758/1959
PWM 653–97653	St Modwen Properties, Long Marston	RH 431760/1959
PWM 654–97654	Peak Rail	RH 431761/1959

▶ The prototype Deltic is seen at NRM Shildon on 5 August 2009 next to 1847 LNWR 2-2-2 3020 "CORNWALL". **Robert Pritchard**

3. ELECTRIC LOCOMOTIVES

Electric railways have existed in Great Britain for over one hundred years. Prior to World War II the majority of electrification was for the movement of passengers in and to metropolitan areas. The North Eastern Railway did however build a small fleet of electric locomotives for hauling heavy coal and steel trains in County Durham.

For notes on wheel arrangements, dimensions, tractive effort and brakes see "Diesel Locomotives" section.

3.1. PRE-NATIONALISATION DESIGN ELECTRIC LOCOMOTIVES

LSWR Bo

Built: 1898. Siemens design for operation on the Waterloo & City line.
System: 750 V DC third rail. **Train Heating:** None.
Traction Motors: Two Siemens 45 kW (60 hp).
Wheel Diameter: 3′ 4″.

BR	*SR*		
DS75	75S	National Railway Museum, Shildon (N)	SM 6/1898

NORTH EASTERN RAILWAY CLASS ES1 Bo-Bo

Built: 1905. 2 built. Used on Newcastle Riverside Branch.
System: 600V DC overhead. **Train Heating:** None.
Traction Motors: 4 BTH design.
Weight: 46 tonnes. **Wheel Diameter:** 915 mm.

BR	*LNER*	*NER*		
26500	1–4075–6480	1	National Railway Museum, Shildon (N)	BE 1905

▲ LSWR 75S at the National Railway Museum, Shildon on 5 August 2009. **Robert Pritchard**

LNER/BR CLASS EM1 (BR CLASS 76) Bo+Bo

Built: 1941–53 at Doncaster (26000) and Gorton (others) for Manchester–Sheffield/Wath system. 58 built.
System: 1500V DC overhead.
Traction Motors: 4 MV 186 axle-hung.
Max Rail Power: 2460 kW (3300 hp).
Continuous Rating: 970 kW (1300 hp).
Maximum Tractive Effort: 200 kN (45 000 lbf).
Continuous Tractive Effort: 39 kN (8800 lbf) at 56 mph.
Weight: 88 tonnes. **Wheel Diameter:** 1270 mm.
Maximum Speed: 65 mph. **Train Heating:** Steam.

26020–E26020–76020	National Railway Museum, York (N)	Gorton 1027/1951

BR CLASS EM2 (BR CLASS 77) Co-Co

Built: 1953–55 at Gorton for BR to LNER design for Manchester–Sheffield/Wath route. 7 built. Sold to NS (Netherlands Railways) 1969.
System: 1500V DC overhead.
Traction Motors: 6 MV 146 axle-hung.
Max Rail Power: 1716 kW (2300 hp).
Maximum Tractive Effort: 200 kN (45 000 lbf).
Continuous Tractive Effort: 78 kN (15 600 lbf) at 23 mph.
Weight: 102 tonnes. **Wheel Diameter:** 1092 mm.
Maximum Speed: 90 mph.
Train Heating: Steam whilst on BR, electric fitted by NS.

Air brakes.

NS	*BR*			
1502	27000–E 27000	ELECTRA	Midland Railway-Butterley	Gorton 1065/1953
1505	27001–E 27001	ARIADNE	Museum of Science & Industry, Manchester	Gorton 1066/1954
1501	27003–E 27003	(DIANA)	Leidschendamm, Den Haag (NS)	Gorton 1068/1954

3.2. BRITISH RAILWAYS ELECTRIC LOCOMOTIVES

NUMBERING SYSTEM

Numbering of electric locomotives from 1957 was similar to that of diesel locomotives, except that the numbers were prefixed with an "E" instead of a "D". Locomotives of pre-nationalisation design continued to be numbered in the 2xxxx series, although Classes EM1 and EM2 later acquired an "E" prefix to their existing numbers. As with diesels, electric locomotives were later allocated a two-digit class number followed by a three-digit serial number.

CLASS 71 Bo-Bo

Built: 1958–60 at Doncaster. 24 built.
System: 660–750 V DC third rail or overhead.
Continuous Rating: 1715 kW (2300 hp).
Maximum Tractive Effort: 191 kN (43 000 lbf).
Continuous Tractive Effort: 55 kN (12 400 lbf) at 69.6 mph.
Weight: 76.2 tonnes. **Wheel Diameter:** 1219 mm.
Maximum Speed: 90 mph. **Train Heating:** Electric

Dual (air/vacuum) braked.

E5001–71001	National Railway Museum, Shildon (N)	Doncaster 1959

CLASS 73/0 ELECTRO-DIESEL — Bo-Bo

Built: 1962 at Eastleigh. 6 built.
System: 660–750 V DC third rail.
Continuous Rating: Electric 1060 kW (1420 hp).
Maximum Tractive Effort: Electric 187 kN (42 000 lbf). Diesel 152 kN (34 100 lbf).
Continuous Tractive Effort: Diesel 72 kN (16 100 lbf) at 10 mph.
Weight: 76.3 tonnes. **Wheel Diameter**: 1016 mm.
Maximum Speed: 80 mph. **Train Heating**: Electric.

Triple (vacuum, air and electro-pneumatic) braked.

Number	Name	Location	Builder
E 6001–73001–73901		Dean Forest Railway	Eastleigh 1962
E 6002–73002		Dean Forest Railway	Eastleigh 1962
E 6003–73003	Sir Herbert Walker	Swindon & Cricklade Railway	Eastleigh 1962
E 6005–73005	Mid Hants WATERCRESS LINE	Knights Rail Services, Eastleigh Works	Eastleigh 1962
E 6006–73006–73906		Crewe Heritage Centre	Eastleigh 1962

CLASS 73/1 ELECTRO-DIESEL — Bo-Bo

Built: 1965–67 by English Electric at Vulcan Foundry, Newton-le-Willows. 43 built.
System: 660–750 V DC third rail.
Continuous Rating: Electric 1060 kW (1420 hp).
Maximum Tractive Effort: Electric 179 kN (40 000 lbf). Diesel 152 kN (34 100 lbf).
Continuous Tractive Effort: Diesel 60 kN (13 600 lbf) at 11.5 mph.
Weight: 76.8 tonnes. **Wheel Diameter**: 1016 mm.
Maximum Speed: 90 mph. **Train Heating**: Electric.

Triple (vacuum, air and electro-pneumatic) braked.

Number	Name	Location	Builder
E 6007–73101	The Royal Alex'	Avon Valley Railway	EE/VF 3569/E339 1965
E 6009–73103		Marshalls Transport, Pershore Airfield	EE/VF 3571/E341 1965
E 6011–73105	Quadrant	Battlefield Railway	EE/VF 3573/E343 1966
E 6016–73110		Nottingham Transport Heritage Centre	EE/VF 3578/E348 1966
E 6019–73113–73211	County of West Sussex	Stewarts Lane Depot, London	EE/VF 3581/E351 1966
E 6020–73114	Stewarts Lane Traction & Maintenance Depot	Battlefield Railway	EE/VF 3582/E352 1966
E 6022–73116–73210	Selhurst	Mid Norfolk Railway	EE/VF 3584/E354 1966
E 6023–73117	University of Surrey	Barrow Hill Roundhouse	EE/VF 3585/E355 1966
E 6024–73118	The Romney, Hythe and Dymchurch Railway	Barry Rail Centre	EE/VF 3586/E356 1966
E 6035–73128	OVS BULLEID C.B.E. 1917–1949 C.M.E Southern Railway	Pontypool & Blaenavon Railway	EE/VF 3597/E367 1966
E 6036–73129	City of Winchester	Gloucestershire Warwickshire Railway	EE/VF 3598/E368 1966
E 6037–73130	City of Portsmouth	Finmere Station, Oxfordshire	EE/VF 3709/E369 1966
E 6041–73134	Woking Homes 1885–1995	Barrow Hill Roundhouse	EE/VF 3713/E373 1966
E 6043–73136	"Perseverance"	Bluebell Railway	EE/VF 3715/E375 1966
E 6047–73140		Spa Valley Railway	EE/VF 3719/E379 1966

73101 also carried the name Brighton Evening Argus.
73136 also carried the name Kent Youth Music.

73101 was numbered 73801 for a time.

CLASS 81 Bo-Bo

Built: 1959–64 by the Birmingham Railway Carriage & Wagon Company, Birmingham. 25 built.
System: 25 kV AC overhead.
Continuous Rating: 2390 kW (3200 hp).
Maximum Tractive Effort: 222 kN (50 000 lbf).
Continuous Tractive Effort: 76 kN (17 000 lbf) at 71 mph.
Weight: 79 tonnes. **Wheel Diameter:** 1219 mm.
Maximum Speed: 100 mph. **Train Heating:** Electric.

Dual (air/vacuum) braked.

E3003–81002 Barrow Hill Roundhouse BTH 1085/1960

CLASS 82 Bo-Bo

Built: 1960–62 by Beyer Peacock, Manchester. 10 built.
System: 25 kV AC overhead.
Continuous Rating: 2460 kW (3300 hp).
Maximum Tractive Effort: 222 kN (50 000 lbf).
Continuous Tractive Effort:76 kN (17 000 lbf) at 73 mph.
Weight: 80 tonnes. **Wheel Diameter:** 1219 mm.
Maximum Speed: 100 mph. **Train Heating:** Electric.

Dual (air/vacuum) braked.

E3054–82008 Barrow Hill Roundhouse BP 7893/1961

CLASS 83 Bo-Bo

Built: 1960–62 by English Electric at Vulcan Foundry, Newton-le-Willows. 15 built.
System: 25 kV AC overhead.
Continuous Rating: 2200 kW (2950 hp).
Maximum Tractive Effort: 169 kN (38 000 lbf).
Continuous Tractive Effort: 68 kN (15 260 lbf) at 73 mph.
Weight: 76 tonnes. **Wheel Diameter:** 1219 mm.
Maximum Speed: 100 mph. **Train Heating:** Electric.

Dual (air/vacuum) braked.

E3035–83012 Barrow Hill Roundhouse EE 2941/VF E277/1961

CLASS 84 Bo-Bo

Built: 1960–61 by North British Locomotive Company, Glasgow. 10 built.
System: 25 kV AC overhead.
Continuous Rating: 2312 kW (3100 hp).
Maximum Tractive Effort: 222 kN (50 000 lbf).
Continuous Tractive Effort: 78 kN (17 600 lbf) at 66 mph.
Weight: 76.6 tonnes. **Wheel Diameter:** 1219 mm.
Maximum Speed: 100 mph. **Train Heating:** Electric.

Dual (air/vacuum) braked.

E3036–84001 Barrow Hill Roundhouse (N) NBL 27793/1960

▲ 73210 "Selhurst" leaves Wymondham on the Mid Norfolk Railway on 21 May 2011, hauling 3 Cig EMU 1497 as the 17.30 Wymondham Abbey–Dereham. **Aubrey Evans**

▼ The AC Locomotive Group has a number of locomotives based at Barrow Hill Roundhouse. On 9 October 2010 89001 and 84001 sandwich 82008. **Paul Abell**

▲ 86259 "Les Ross" passes Red Bank at Warrington with a Vintage Trains railtour from Carnforth to Tyseley on 30 May 2009. **Andrew Mason**

▼ 87002 "The AC Locomotive Group" passes Grassthorpe (between Retford and Newark) on a Doncaster–Brighton railtour on 25 July 2009. **Andrew Mason**

CLASS 85 Bo-Bo

Built: 1961–65 at Doncaster. 40 built.
System: 25 kV AC overhead.
Continuous Rating: 2390 kW (3200 hp).
Maximum Tractive Effort: 222 kN (50 000 lbf).
Continuous Tractive Effort: 76 kN (17 000 lbf) at 71 mph.
Weight: 82.5 tonnes. **Wheel Diameter:** 1219 mm.
Maximum Speed: 100 mph. **Train Heating:** Electric.

Dual (air/vacuum) braked.

E3061–85006–85101	"Doncaster Plant 150 1853–2003"	Barrow Hill Roundhouse	Doncaster 1961

CLASS 86 Bo-Bo

Built: 1965–66 at Doncaster or English Electric Company at Vulcan Foundry, Newton-le-Willows. 100 built.
System: 25 kV AC overhead.
Continuous Rating: 2680 kW (3600 hp) (* 5860 kW (5000 hp)) († 3010 kW (4040 hp)).
Maximum Tractive Effort: 258 kN (58 000 lbf) († 207 kN (46 500 lbf).
Continuous Tractive Effort: 89 kN (20 000 lbf) (* 95 kN 21 300 lbf) († 85 kN 19 200 lbf).
Weight: 83 tonnes (* 87 tonnes) († 85 tonnes). **Wheel Diameter:** 1156 mm (* 1150 mm).
Maximum Speed: 100 mph (* 110 mph). **Train Heating:** Electric.

Dual (air/vacuum) braked.

E3137–86045–86259†	"Les Ross"	Tyseley Locomotive Works	Doncaster 1966
E3191–86201–86101*	Sir William A Stanier FRS	LNWR, Crewe Carriage Shed	EE/VF3483/E329 1965
E3193–86213†	Lancashire Witch	Wembley Depot, London	EE/VF3485/E331 1965
E3199–86001–86401	Northampton Town	Willesden Depot, London	EE/VF3491/E337 1966

86259 also carried the names Peter Pan and Greater MANCHESTER THE LIFE & SOUL OF BRITAIN.

CLASS 87 Bo-Bo

Built: 1973–75 by BREL at Crewe. 36 built.
System: 25 kV AC overhead.
Continuous Rating: 3730 kW (5000 hp).
Maximum Tractive Effort: 258 kN (58 000 lbf).
Continuous Tractive Effort: 95 kN (21 300 lbf) at 87 mph.
Weight: 83.5 tonnes. **Wheel Diameter:** 1150 mm.
Maximum Speed: 110 mph. **Train Heating:** Electric.

Air braked

87001	Royal Scot	National Railway Museum, York (N)	Crewe 1973
87002	"The AC Locomotive Group"	LNWR, Crewe Carriage Shed	Crewe 1973
87035	Robert Burns	Crewe Heritage Centre	Crewe 1974

87001 also carried the name STEPHENSON and 87002 was originally named Royal Sovereign.

CLASS 89 Co-Co

Built: 1987 by BREL at Crewe. 1 built.
System: 25 kV AC overhead.
Continuous Rating: 4350 kW (6550 hp).
Maximum Tractive Effort: 205 kN (46 000 lbf).
Continuous Tractive Effort: 105 kN (23 600 lbf) at 92 mph.
Weight: 104 tonnes. **Wheel Diameter:** 1150 mm.
Maximum Speed: 125 mph. **Train Heating:** Electric.

Air braked

89001	Avocet	Barrow Hill Roundhouse	Crewe 1987

4. GAS TURBINE VEHICLES

LOCOMOTIVE — A1A-A1A

Built: 1950 by Brown Boveri in Switzerland.
Power Unit: Brown Boveri gas turbine of 1828 kW (2450 hp).
Transmission: Electric. Four traction motors.
Maximum Tractive Effort: 140 kN (31 500 lbf).
Continuous Tractive Effort: 55 kN (12 400 lbf) at 64 mph.
Weight: 117.1 tonnes.
Wheel Diameter: 1232 mm.
Maximum Speed: 90 mph.
Train Heating: Steam.

18000	Gloucestershire Warwickshire Railway	BBC 4559/1950

EXPERIMENTAL ADVANCED PASSENGER TRAIN (APT-E)

Built: 1972 at Derby Litchurch Lane.
Power Units: Eight Leyland 350 automotive gas turbines of 222 kW (298 hp).
Traction Motors: Four GEC 253AY. Articulated unit.

PC1	National Railway Museum, Shildon (N)	Derby 1972
PC2	National Railway Museum, Shildon (N)	Derby 1972
TC1	National Railway Museum, Shildon (N)	Derby 1972
TC2	National Railway Museum, Shildon (N)	Derby 1972

▲ **After a light dusting of snow, unique 89001 is being shunted by 47841 at Barrow Hill Roundhouse on 28 November 2010.** **Phil Chilton**

5. MULTIPLE UNIT VEHICLES

GENERAL

Prior to nationalisation, several schemes to transfer bus technology to rail vehicles had taken place with little success, the exception being the GWR where the concept was developed resulting in a fleet of distinctive diesel railcars.

During the 1950s British Railways took the idea far more seriously as part of the modernisation plan and consequently numerous designs appeared. The preservation of these has expanded enormously in recent years.

Whilst the majority of units in this section are diesel-powered, the first two items which qualify for inclusion are their predecessors in the form of the recently restored Steam Railmotor, followed by the Petrol-Electric Autocar currently undergoing restoration.

TYPE CODES

The type codes used by the former BR operating departments to describe the various types of multiple unit vehicles are used, these being:

B Brake, ie a vehicle with luggage space and a guard's/conductor's compartment.

BDM	Battery driving motor	M	Motor
BDT	Battery driving trailer	O	Open vehicle
C	Composite	P	Pullman
Cso	Semi-open composite	PMV	Parcels & miscellaneous van
DM	Driving motor	RSB	Buffet car with Second Class seating
DT	Driving trailer	RSKB	Kitchen/buffet car with Second Class seating
F	First	S	Second (now known as Standard)
K	Side corridor with lavatory or kitchen	Sso	Semi-open second
L	Open or semi-open with lavatory	T	Third (reclassified second in 1956)
LV	Luggage Van	T	Trailer

All diesel mechanical and diesel hydraulic vehicles are assumed to be open unless stated otherwise and do not carry an "O" in the code. Some DMU vehicles are usually used as hauled stock and these are denoted by a letter "h" after the number.

PREFIXES AND SUFFIXES

Coaching stock vehicles used to carry regional prefix letters to denote the owning region. These were removed in the 1980s. These are not shown. Pre-nationalisation number series vehicles carried both prefix and suffix letters, the suffix denoting the pre-nationalisation number series. The prefixes and suffixes are shown for these vehicles.

DIMENSIONS

Dimensions are shown as length (over buffers or couplers) x width (over bodysides including door handles).

SEATING CAPACITIES

These are shown as nF/nS relating to First and Second Class seats respectively, eg a car with 12 First Class seats and 51 Second Class seats would be shown as 12/51. Prior to 3 June 1956 Second class was referred to as "Third" Class and is now referred to as "Standard" class. Certain old vehicles are thus shown as Third Class.

BOGIES

All vehicles are assumed to have two four-wheeled bogies unless otherwise stated.

5.1. STEAM RAILMOTOR

Original Class Built: 1905–08. 35 built. Converted to Autotrailer 212 in 1935. Converted back to Steam Railmotor in 2011 with new replica steam propulsion unit.
Boiler Pressure: 160 lbf/sq in. **Weight:** 45.55 tons.
Wheel Diameters: 4′ 0″ (driving), 3′ 7½″. **Valve Gear:** Walschaerts.
Cylinders: 12″ x 16″ (O).
Tractive Effort: 6530 lbf.
Seats: –/61.

93 Didcot Railway Centre Rebuilt Llangollen 2011

5.2 PETROL-ELECTRIC AUTOCAR

Built as a petrol-electric autocar for the North Eastern Railway, this vehicle is now being restored to its rebuilt 1923 form as a diesel railcar using a motor coach from a Class 416 EMU. It will operate with NER trailer autocoach 3453.
Body is original chassis from LNER 60525. Mechanical parts from BR EMU including power bogie.
Original Class Built: York 1903 (2 built). **Weight:** 35 tons.
Transmission: . **Wheel Diameters:** .
Engine: New engine being built to 225 hp.
Seats: –/48 (–/52 as built).
To be fitted with vacuum and air brakes.

3170 Embsay & Bolton Abbey Railway Undergoing rebuilding

5.3. GWR DIESEL RAILCARS

UNCLASSIFIED PARK ROYAL

Built: 1934 by Park Royal. Single cars with two driving cabs.
Engines: Two AEC 90 kW (121 hp). **Transmission:** Mechanical.
Body: 19.58 x 2.70 m. **Weight:** 26.6 tonnes.
Seats: –/44.
Max. Speed: 75 mph.

BR	*GWR*		
W4W	4	Steam – Museum of the Great Western Railway, Swindon (N)	PR 1934

UNCLASSIFIED GWR

Built: 1940 at Swindon. Single cars with two driving cabs.
Engines: Two AEC 78 kW (105 hp). **Transmission:** Mechanical.
Body: 20.21 x 2.70 m. **Weight:** 36.2 tonnes.
Seats: –/48.
Max. Speed: 40 mph.

BR	*GWR*		
W 20 W	20	Kent & East Sussex Railway	Swindon 1940
W 22 W	22	Didcot Railway Centre	Swindon 1941

▲ Superbly restored Steam Railmotor No. 93 on display at Didcot Railway Centre on 29 May 2011.
Brian Garvin

▼ Also at Didcot is GWR Railcar No. 22, seen on 24 May 2011. **Paul Abell**

5.4. BRITISH RAILWAYS DMUs

NUMBERING SYSTEM

Early BR Diesel Multiple Units were numbered in the 79xxx series, but when it was evident that this series did not contain enough numbers the 5xxxx series was allocated to this type of vehicle and the few locomotive-hauled non-corridor coaches which were in the 5xxxx series were renumbered into the 4xxxx series. Power cars in the 50xxx series and driving trailers in the 56xxx series were eventually renumbered into the 53xxx and 54xxx series respectively to avoid conflicting numbers with Class 50 and 56 diesel locomotives.

Diesel Electric Multiple Unit power cars were renumbered in the 60000–60499 series, trailers in the 60500–60799 series and driving trailers in the 60800–60999 series.

5.4.1. HIGH SPEED DIESEL TRAINS

The prototype High Speed Diesel Train (HSDT) appeared in 1972. The two power cars (41001/002) were constructed to a locomotive lot (No. 1501) whilst the intermediate vehicles (10000, 10100, 11000–11002 and 12000–12002) were constructed to coaching stock lots. On 10 July 1974 the power cars were reclassified as coaching stock and issued with a coaching stock lot number. All prototype HSDT vehicles were thus categorised as multiple unit stock and renumbered into the 4xxxx series, Class No. 252 being allocated for the complete set.

CLASS 252 — PROTOTYPE HST POWER CAR

Built: 1972 at Derby. 2 built.
Engine: Paxman Valenta 12RP200L of 1680 kW (2250 hp) at 1500 rpm.
Traction Motors: Four Brush TMH 68-46.
Power at Rail: 1320 kW (1770 hp).
Maximum Tractive Effort: 80 kN (17 980 lbf).
Continuous Tractive Effort: 46 kN (10 340 lbf) at 64.5 mph.
Weight: 67 tonnes. **Wheel Diameter:** 1020 mm.
Maximum Speed: 125 mph.

Air braked.

41001–43000–ADB 975812	National Railway Museum, York (N)	Derby 1972

5.4.2. DIESEL MECHANICAL MULTIPLE UNITS

All vehicles in this section have a Maximum Speed of 70 mph.

CLASS 100 — GRCW 2-CAR UNITS

Built: 1957–58. Normal formation: DMBS–DTCL.
Engines: Two AEC 220 of 112 kW (150 hp).

DMBS	18.49 x 2.82 m	30.5 tonnes	–/52
DTCL	18.49 x 2.82 m	25.5 tonnes	12/54

51118	DMBS	Midland Railway-Butterley	GRCW 1957
56097	DTCL	Midland Railway-Butterley	GRCW 1957
56301	DTCL	Mid Norfolk Railway	GRCW 1957
56317	DTCL	North Essex Traction Group	GRCW 1958

CLASS 101 METRO-CAMMELL UNITS

Built: 1958–59. Various formations.
Engines: Two AEC 220 of 112 kW (150 hp).

DMBS	18.49 x 2.82 m	32.5 tonnes	–/52
DMCL	18.49 x 2.82 m	32.5 tonnes	12/46 (originally 12/53)
DTCL	18.49 x 2.82 m	25.5 tonnes	12/53
TCL	18.49 x 2.82 m	25.5 tonnes	12/53
TSL	18.49 x 2.82 m	25.5 tonnes	–/71

50160–53160	DMCL	Midland Railway-Butterley	MC 1956
50164–53164	DMBS	Midland Railway-Butterley	MC 1956
50167–53167–977392	DMBS	Churnet Valley Railway	MC 1956
50170–53170	DMCL	Ecclesbourne Valley Railway	MC 1957
50193–53193–977898	DMCL	Great Central Railway	MC 1957
50203–53203–977897	DMBS	Great Central Railway	MC 1957
50204–53204	DMBS	North Yorkshire Moors Railway	MC 1957
50222–53222–977693	DMBS	Barry Rail Centre	MC 1957
50253–53253	DMBS	Midland Railway-Butterley	MC 1957
50256–53256	DMBS	East Kent Light Railway	MC 1957
50266–53266	DMCL	Great Central Railway	MC 1957
50321–53321–977900	DMCL	Great Central Railway	MC 1958
50338–53338–977694	DMCL	Barry Rail Centre	MC 1958
50746–53746	DMCL	Wensleydale Railway	MC 1957
51187	DMBS	Cambrian Railway Trust, Llynclys	MC 1958
51188	DMBS	Ecclesbourne Valley Railway	MC 1958
51189	DMBS	Keighley & Worth Valley Railway	MC 1958
51192	DMBS	East Lancashire Railway (N)	MC 1958
51205	DMBS	Cambrian Railway Trust, Llynclys	MC 1958
51210	DMBS	Wensleydale Railway	MC 1958
51213	DMBS	East Anglian Railway Museum	MC 1958
51226	DMBS	Mid Norfolk Railway	MC 1958
51228	DMBS	North Norfolk Railway	MC 1958
51247	DMBS	Wensleydale Railway	MC 1958
51427–977899	DMBS	Great Central Railway	MC 1959
51433–977391	DMBS	Churnet Valley Railway	MC 1959
51434 "Matthew Smith"	DMBS	Mid Norfolk Railway	MC 1959
51499	DMCL	Mid Norfolk Railway	MC 1959
51503	DMCL	Mid Norfolk Railway	MC 1959
51505	DMCL	Ecclesbourne Valley Railway	MC 1959
51511	DMCL	North Yorkshire Moors Railway	MC 1959
51512	DMCL	Cambrian Railway Trust, Llynclys	MC 1959
51803	DMCL	Keighley & Worth Valley Railway	MC 1959
56055–54055	DTCL	Cambrian Railway Trust, Llynclys	MC 1957
56062–54062	DTCL	North Norfolk Railway	MC 1957
56342–54342–042222	DTCL	Midland Railway-Butterley	MC 1958
56343–54343	DTCL	East Kent Light Railway	MC 1958
56347–54347	DTCL	Bressingham Steam Museum	MC 1958
56352–54352	DTCL	East Lancashire Railway (N)	MC 1958
56356–54356–6300h	DTCL	Gloucestershire Warwickshire Railway	MC 1959
56358–54358	DTCL	East Anglian Railway Museum	MC 1959
56365–54365	DTCL	East Anglian Railway Museum	MC 1958
56408	DTCL	Spa Valley Railway	MC 1958
59117	TCL	Mid Norfolk Railway	MC 1958
59303	TSL	Ecclesbourne Valley Railway	MC 1957
59539	TCL	North Yorkshire Moors Railway	MC 1959

▲ North Yorkshire Moors Railway Class 101 DMU 50204+59539+51511 passes Green End whilst working the 15.30 Grosmont–Goathland on 20 September 2009. **Lindsay Atkinson**

▼ Some railways, in particular the Dartmouth Steam Railway and the West Somerset Railway, use DMU trailers as hauled stock. On 16 August 2010 Class 116 TS 59004 "EMMA" forms part of a train from Paignton to Kingswear. **Robert Pritchard**

CLASS 103 PARK ROYAL 2-CAR UNITS

Built: 1958. Normal formation: DMBS–DTCL.
Engines: Two AEC 220 of 112kW (150 hp).

DMBS	18.49 x 2.82 m	34 tonnes	–/52
DTCL	18.49 x 2.82 m	27 tonnes	16/48

50413	DMBS	Helston Railway	PR 1958
56160–DB975228	DTCL	Denbigh & Mold Junction Railway	PR 1958
56169	DTCL	Helston Railway	PR 1958

CLASS 104 BRCW UNITS

Built: 1957–58. Various formations.
Engines: Two BUT (Leyland) of 112 kW (150 hp).

DMBS	18.49 x 2.82 m	31.5 tonnes	–/52
TCL	18.49 x 2.82 m	24.5 tonnes	12/54
TBSL	18.49 x 2.82 m	25.5 tonnes	–/51
DMCL	18.49 x 2.82 m	31.5 tonnes	12/54 (* 12/51)
DTCL	18.49 x 2.82 m	24.5 tonnes	12/54

50437–53437	DMBS	Churnet Valley Railway	BRCW 1957
50447–53447	DMBS	Llangollen Railway	BRCW 1957
50454–53454	DMBS	Llangollen Railway	BRCW 1957
50455–53455	DMBS	Churnet Valley Railway	BRCW 1957
50479–53479	DMBS	Telford Steam Railway	BRCW 1958
50494–53494	DMCL	Churnet Valley Railway	BRCW 1957
50517–53517	DMCL	Churnet Valley Railway	BRCW 1957
50528–53528	DMCL	Llangollen Railway	BRCW 1958
50531–53531	DMCL	Telford Steam Railway	BRCW 1958
50556–53556*	DMCL	Telford Steam Railway	BRCW 1958
56182–54182–977554	DTCL	Churnet Valley Railway	BRCW 1958
59137	TCL	Churnet Valley Railway	BRCW 1957
59228	TBSL	Telford Steam Railway	BRCW 1958

CLASS 105 CRAVEN 2-CAR UNITS

Built: 1957–59. Various formations: DMBS–DTCL or DMCL.
Engines: Two AEC 220 of 112 kW (150 hp).

DMBS	18.49 x 2.82 m	29.5 tonnes	–/52
DTCL	18.49 x 2.82 m	23.5 tonnes	12/51

51485	DMBS	East Lancashire Railway	Cravens 1959
56121	DTCL	East Lancashire Railway	Cravens 1957
56456–54456	DTCL	Llangollen Railway	Cravens 1959

CLASS 107 DERBY HEAVYWEIGHT 3-CAR UNITS

Built: 1960–61. Normal formation: DMBS–TSL–DMCL.
Engines: Two BUT (Leyland) of 112 kW (150 hp).

DMBS	18.49 x 2.82 m	35 tonnes	–/52
DMCL	18.49 x 2.82 m	35.5 tonnes	12/53
TSL	18.49 x 2.82 m	28.5 tonnes	–/71

51990–977830	DMBS	Strathspey Railway	Derby 1960
51993–977834	DMBS	Tanat Valley Light Railway	Derby 1961
52005–977832	DMBS	Nene Valley Railway	Derby 1961
52006	DMBS	Avon Valley Railway	Derby 1961
52008	DMBS	Strathspey Railway	Derby 1961
52012–977835	DMCL	Tanat Valley Light Railway	Derby 1960
52025–977833	DMCL	Avon Valley Railway	Derby 1961
52029 h	DMCL	Llanelli & Mynydd Mawr Railway	Derby 1961
52030–977831	DMBS	Strathspey Railway	Derby 1961
52031	DMCL	Nene Valley Railway	Derby 1961
59791	TSL	Nene Valley Railway	Derby 1961

CLASS 108 — DERBY LIGHTWEIGHT UNITS

Built: 1958–1961. Various formations. **Engines:** Two Leyland of 112 kW (150 hp).

DMBS	18.49 x 2.79 m	29.5 tonnes	–/52
TBSL	18.49 x 2.79 m	21.5 tonnes	–/50
TSL	18.49 x 2.79 m	21.5 tonnes	–/68
DMCL	18.49 x 2.79 m	28.5 tonnes	12/53
DTCL	18.49 x 2.79 m	21.5 tonnes	12/53

50599–53599	DMBS	Ecclesbourne Valley Railway	Derby 1958
50619–53619	DMBS	Dean Forest Railway	Derby 1958
50628–53628	DMBS	Keith & Dufftown Railway	Derby 1958
50632–53632	DMCL	Pontypool & Blaenavon Railway	Derby 1958
50645–53645	DMCL	Nottingham Transport Heritage Centre	Derby 1958
50926–53926–977814	DMBS	Nottingham Transport Heritage Centre	Derby 1959
50928–53928	DMBS	Keighley & Worth Valley Railway	Derby 1959
50933–53933	DMBS	Severn Valley Railway	Derby 1960
50971–53971	DMBS	Kent & East Sussex Railway	Derby 1959
50980–53980	DMBS	Bodmin & Wenford Railway	Derby 1959
51562	DMCL	East Lancashire Railway (N)	Derby 1959
51565	DMCL	Keighley & Worth Valley Railway	Derby 1959
51566	DMCL	Dean Forest Railway	Derby 1959
51567-977854	DMCL	Midland Railway-Butterley	Derby 1959
51568	DMCL	Keith & Dufftown Railway	Derby 1959
51571	DMCL	Kent & East Sussex Railway	Derby 1960
51572	DMCL	Stainmore Railway	Derby 1960
51907	DMBS	Llangollen Railway	Derby 1960
51909	DMBS	St Modwen Properties, Long Marston	Derby 1960
51914	DMBS	Dean Forest Railway	Derby 1960
51919	DMBS	Garw Valley Railway	Derby 1960
51922	DMBS	East Lancashire Railway (N)	Derby 1960
51933	DMBS	Swanage Railway	Derby 1960
51937–977806	DMBS	Midland Railway-Butterley	Derby 1960
51941	DMBS	Severn Valley Railway	Derby 1960
51942	DMBS	Pontypool & Blaenavon Railway	Derby 1961
51947	DMBS	Bodmin & Wenford Railway	Derby 1961
51950	DMBS	Gloucestershire Warwickshire Railway	Derby 1961
52044	DMCL	Pontypool & Blaenavon Railway	Derby 1960
52048	DMCL	Garw Valley Railway	Derby 1960
52053–977807	DMCL	Keith & Dufftown Railway	Derby 1960
52054	DMCL	Bodmin & Wenford Railway	Derby 1960
52062	DMCL	Gloucestershire Warwickshire Railway	Derby 1961
52064	DMCL	Severn Valley Railway	Derby 1961
56207–54207 h	DTCL	Appleby-Frodingham RPS, Scunthorpe	Derby 1958
56208–54208	DTCL	Severn Valley Railway	Derby 1958
56223–54223	DTCL	Llangollen Railway	Derby 1959
56224–54224	DTCL	Keith & Dufftown Railway	Derby 1959
56270–54270	DTCL	Pontypool & Blaenavon Railway	Derby 1959
56271–54271	DTCL	St Modwen Properties, Long Marston	Derby 1960
56274–54274 h	DTCL	Stainmore Railway	Derby 1960
56279–54279	DTCL	Lavender Line	Derby 1960
56484–54484	DTCL	Midland Railway-Butterley	Derby 1960
56490–54490	DTCL	Llangollen Railway	Derby 1960
56491–54491	DTCL	Keith & Dufftown Railway	Derby 1960
56492–54492	DTCL	Dean Forest Railway	Derby 1960
56495–54495	DTCL	Kirklees Light Railway	Derby 1960
56504–54504	DTCL	Swanage Railway	Derby 1960
59245 h	TBSL	Appleby-Frodingham RPS, Scunthorpe	Derby 1958
59250 "THE CHATHAM BAR"	TBSL	Severn Valley Railway	Derby 1958
59387	TSL	Dean Forest Railway	Derby 1958

50628 and 56491 are named "SPIRIT OF DUFFTOWN" and 51568 & 52053 are named "SPIRIT OF BANFFSHIRE".

CLASS 109 D. WICKHAM 2-CAR UNITS

Built: 1957. Normal formation: DMBS–DTCL.
Engines: Two BUT (Leyland) of 112 kW (150 hp).

DMBS	18.49 x 2.82 m	27.5 tonnes	–/52
DTCL	18.49 x 2.82 m	20.5 tonnes	16/50

50416–DB 975005	DMBS	Llangollen Railway	Wkm 1957
56171–DB 975006	DTCL	Llangollen Railway	Wkm 1957

CLASS 110 BRCW CALDER VALLEY 3-CAR UNITS

Built: 1961–62. Normal formation: DMBC–TSL–DMCL.
Engines: Two Rolls-Royce C6NFLH38D of 134 kW (180 hp).

DMBC	18.48 x 2.82 m	32.5 tonnes	12/33
DMCL	18.48 x 2.82 m	32.5 tonnes	12/54
TSL	18.48 x 2.82 m	24.5 tonnes	–/72

51813	DMBC	Wensleydale Railway	BRCW 1961
51842	DMCL	Wensleydale Railway	BRCW 1961
52071	DMBC	Lakeside & Haverthwaite Railway	BRCW 1962
52077	DMCL	Lakeside & Haverthwaite Railway	BRCW 1961
59701	TSL	Churnet Valley Railway	BRCW 1961

59701 is on loan from the Wensleydale Railway.

CLASS 111 METRO-CAMMELL TRAILER BUFFET

Built: 1960. Used to augment other units as required.

TRSBL	18.49 x 2.82 m	25.5 tonnes	–/53

59575	TRSBL	Great Central Railway	MC 1960

CLASS 114 DERBY HEAVYWEIGHT 2-CAR UNITS

Built: 1956–57. Normal formation: DMBS–DTCL.
Engines: Two Leyland Albion of 149 kW (200 hp).

DMBS	20.45 x 2.82 m	38 tonnes	–/62
DTCL	20.45 x 2.82 m	30 tonnes	12/62

50015–53015–55929–977775	DMBS	Midland Railway-Butterley	Derby 1957
50019–53019	DMBS	Midland Railway-Butterley	Derby 1957
56006–54006	DTCL	Midland Railway-Butterley	Derby 1956
56015–54015–54904–977776	DTCL	Midland Railway-Butterley	Derby 1957
56047–54047	DTCL	Strathspey Railway	Derby 1957

CLASS 115 DERBY SUBURBAN 4-CAR UNITS

Built: 1960. Non-gangwayed when built, but gangways subsequently fitted. Normal formation: DMBS–TSso–TCL–DMBS.
Engines: Two Leyland Albion of 149 kW (200 hp).

DMBS	20.45 x 2.82 m	38.5 tonnes	–/74 (originally –/78)
TCL	20.45 x 2.82 m	30.5 tonnes	28/38 (originally 30/40)
TSso	20.45 x 2.82 m	29.5 tonnes	–/98 (originally –/106)

59678 has been converted to TCLRB seating 28/32.

BR	*Present*			
51655		DMBS	Barry Rail Centre	Derby 1960
51669		DMBS	Spa Valley Railway	Derby 1960
51677		DMBS	Barry Rail Centre	Derby 1960
51849		DMBS	Spa Valley Railway	Derby 1960
51852		DMBS	West Somerset Railway	Derby 1960
51859		DMBS	West Somerset Railway	Derby 1960

51880	DMBS	West Somerset Railway	Derby 1960
51886	DMBS	Buckinghamshire Railway Centre	Derby 1960
51887	DMBS	West Somerset Railway	Derby 1960
51899	DMBS	Buckinghamshire Railway Centre	Derby 1960
59659 h "9659"	TSso	Midland Railway-Butterley	Derby 1960
59664	TCL	Barry Rail Centre	Derby 1960
59678	TCL	West Somerset Railway	Derby 1960
59719	TCL	South Devon Railway	Derby 1960
59740 h "9740"	TSso	South Devon Railway	Derby 1960
59761	TCL	Buckinghamshire Railway Centre	Derby 1960

51899 is named "AYLESBURY COLLEGE SILVER JUBILEE".

CLASS 116 DERBY SUBURBAN 3-CAR UNITS

Built: 1957–58. Non-gangwayed when built, but gangways subsequently fitted. Normal formation: DMBS–TS or TC–DMS.
Engines: Two Leyland of 112 kW (150 hp).

DMBS	20.45 x 2.82 m	36.5 tonnes	–/65
DMS	20.45 x 2.82 m	36.5 tonnes	–/89 (originally –/95)
TS§	20.45 x 2.82 m	29 tonnes	–/98 (originally –/102)
TC	20.45 x 2.82 m	29 tonnes	20/68 (originally 28/74)

§-converted from TC seating 28/74.

51131	DMBS	Battlefield Railway	Derby 1958
51138–977921	DMBS	Nottingham Transport Heritage Centre	Derby 1958
51151	DMS	Nottingham Transport Heritage Centre	Derby 1958
59003 h "ZOE"	TS	Dartmouth Steam Railway	Derby 1957
59004 h "EMMA"	TS	Dartmouth Steam Railway	Derby 1957
59444 h	TC	Chasewater Light Railway	Derby 1958

CLASS 117 PRESSED STEEL SUBURBAN 3-CAR UNITS

Built: 1960. Non-gangwayed when built, but gangways subsequently fitted. Normal formation: DMBS–TCL–DMS.
Engines: Two Leyland of 112 kW (150 hp).

DMBS	20.45 x 2.82 m	36.5 tonnes	–/65
TCL	20.45 x 2.82 m	30.5 tonnes	22/48 (originally 24/50)
DMS	20.45 x 2.82 m	36.5 tonnes	–/89 (originally –/91)

51339	DMBS	Gloucestershire Warwickshire Railway	PS 1960
51341	DMBS	Midland Railway-Butterley	PS 1960
51342	DMBS	Epping–Ongar Railway	PS 1960
51346	DMBS	Swanage Railway	PS 1960
51347	DMBS	Gwili Railway	PS 1960
51351	DMBS	Pontypool & Blaenavon Railway	PS 1960
51352	DMBS	St Modwen Properties, Long Marston	PS 1960
51353	DMBS	Midland Railway-Butterley	PS 1960
51354	DMBS	Llanelli & Mynydd Mawr Railway	PS 1960
51356	DMBS	Allely's, The Slough, Studley	PS 1960
51359	DMBS	Northampton & Lamport Railway	PS 1960
51360	DMBS	Ecclesbourne Valley Railway	PS 1960
51363	DMBS	Mid Hants Railway	PS 1960
51365	DMBS	Gloucestershire Warwickshire Railway	PS 1960
51367	DMBS	Strathspey Railway	PS 1960
51370	DMBS	Titley Junction Station, Herefordshire	PS 1960
51372	DMBS	Titley Junction Station, Herefordshire	PS 1960
51376	DMS	St Modwen Properties, Long Marston	PS 1960
51381	DMS	Mangapps Railway Museum	PS 1960
51382	DMS	Gloucestershire Warwickshire Railway	PS 1960
51384	DMS	Epping–Ongar Railway	PS 1960
51388	DMS	Swanage Railway	PS 1960
51392	DMS	Allely's, The Slough, Studley	PS 1960

▲ Battlefield Line Class 118/116 DMU (51321+51131) displays all over BR Blue at Shenton on the 15.15 to Shackerstone on 18 April 2009. **Phil Barnes**

▼ Class 122 bubble car 55000 passes Caddaford with the 10.00 Buckfastleigh–Totnes Riverside on 28 August 2010. **Callum Hayes**

51395	DMS	Midland Railway-Butterley	PS 1960
51396	DMS	Llanelli & Mynydd Mawr Railway	PS 1960
51397	DMS	Pontypool & Blaenavon Railway	PS 1960
51398	DMS	Midland Railway-Butterley	PS 1960
51400	DMS	Wensleydale Railway	PS 1960
51401	DMS	Gwili Railway	PS 1960
51402	DMS	Strathspey Railway	PS 1960
51405	DMS	Mid Hants Railway	PS 1960
51407	DMS	Gloucestershire Warwickshire Railway	PS 1960
51412	DMS	Titley Junction Station, Herefordshire	PS 1960
59486	TCL	Swanage Railway	PS 1960
59488	TCL	Dartmouth Steam Railway	PS 1960
59492	TCL	Allely's, The Slough, Studley	PS 1960
59493 h	TCL	West Somerset Railway	PS 1960
59494 h "CHLOE"	TCL	Dartmouth Steam Railway	PS 1960
59500	TCL	Wensleydale Railway	PS 1960
59501	TCL	Nottingham Transport Heritage Centre	PS 1960
59503 h "NINA"	TCL	Dartmouth Steam Railway	PS 1960
59505	TCL	St Modwen Properties, Long Marston	PS 1960
59506 h	TCL	West Somerset Railway	PS 1960
59507 h "ELLA"	TCL	Dartmouth Steam Railway	PS 1960
59508	TCL	Gwili Railway	PS 1960
59509	TCL	Wensleydale Railway	PS 1960
59510	TCL	Mid Hants Railway	PS 1960
59511	TCL	Strathspey Railway	PS 1960
59513 h "HEIDI"	TCL	Dartmouth Steam Railway	PS 1960
59514	TCL	Swindon & Cricklade Railway	PS 1960
59515 h	TCL	West Somerset Railway	PS 1960
59516	TCL	Swanage Railway	PS 1960
59517 h "EMILY"	TCL	Dartmouth Steam Railway	PS 1960
59520	TCL	Dartmoor Railway	PS 1960
59521	TCL	Midland Railway-Butterley	PS 1960
59522 h	TCL	Chasewater Light Railway	PS 1960

51365 and 51407 are on loan from the Plym Valley Railway.

59488 has been converted into a static Visitor Centre at Kingswear Station.

59510 incorrectly carries the number 59515.

CLASS 118 BRCW SUBURBAN 3-CAR UNITS

Built: 1960. Non-gangwayed when built, but gangways subsequently fitted. Normal formation: DMBS–TCL–DMS.
Engines: Two Leyland of 112 kW (150 hp).

DMS	20.45 x 2.82 m	36.5 tonnes	–/89 (originally –/91)

51321–977753	DMS	Battlefield Railway	BRCW 1960

CLASS 119 GRCW CROSS-COUNTRY 3-CAR UNITS

Built: 1959. Normal formation: DMBC–TSLRB–DMSL.
Engines: Two Leyland of 112 kW (150 hp).

DMBC	20.45 x 2.82 m	37.5 tonnes	18/16
DMSL	20.45 x 2.82 m	38.5 tonnes	–/68

51073	DMBC	Ecclesbourne Valley Railway	GRCW 1959
51074	DMBC	Swindon & Cricklade Railway	GRCW 1959
51104	DMSL	Swindon & Cricklade Railway	GRCW 1959

CLASS 120 SWINDON CROSS-COUNTRY 3-CAR UNITS

Built: 1958. Normal formation: DMBC–TRSBL–DMSL.

TRSBL	20.45 x 2.82 m	31.5 tonnes	–/60

59276	TRSBL	Great Central Railway	Swindon 1958

CLASS 121 PRESSED STEEL SINGLE UNITS & DRIVING TRAILERS

Built: 1960–61. Non-gangwayed single cars with two driving cabs plus driving trailers used for augmentation. The driving trailers were latterly fitted with gangways for coupling to power cars of other classes.
Engines: Two Leyland of 112 kW (150 hp).
Transmission: Mechanical.

DMBS	20.45 x 2.82 m	38 tonnes	–/65
DTS	20.45 x 2.82 m	30 tonnes	–/89 (originally –/91)

55023	DMBS	Chinnor & Princes Risborough Railway	PS 1960
55028–977860	DMBS	Swanage Railway	PS 1960
55029–977968	DMBS	Rushden Transport Museum	PS 1960
55033–977826	DMBS	Colne Valley Railway	PS 1960
56287–54287	DTS	Colne Valley Railway	PS 1961
56289–54289	DTS	East Lancashire Railway	PS 1961

CLASS 122 GRCW SINGLE-CAR UNITS

Built: 1958. Non-gangwayed single cars with two driving cabs.
Engines: Two AEC 220 of 112 kW (150 hp).

DMBS	20.45 x 2.82 m	36.5 tonnes	–/65

55000	DMBS	South Devon Railway	GRCW 1958
55001-DB975023	DMBS	East Lancashire Railway	GRCW 1958
55003	DMBS	Gloucestershire Warwickshire Railway	GRCW 1958
55005	DMBS	Battlefield Railway	GRCW 1958
55006	DMBS	Ecclesbourne Valley Railway	GRCW 1958
55009	DMBS	Mid Norfolk Railway	GRCW 1958
55012	DMBS	Rail Restorations North East, Shildon	GRCW 1958

CLASS 126 SWINDON INTER-CITY UNITS

Built: 1956–59. 51017, 51043 and 59404 for Ayrshire services and 79443 for Glasgow–Edinburgh services. Various formations.
Engines: Two AEC 220 of 112 kW (150 hp).

DMBSL	20.45 x 2.82 m	38.5 tonnes	–/52
TCK	20.45 x 2.82 m	32.3 tonnes	18/32
TRCsoKBL	20.45 x 2.82 m	34 tonnes	18/12. First Class seating in compartments.
DMSL	20.45 x 2.82 m	38.5 tonnes	–/64

51017	DMSL	Bo'ness & Kinneil Railway	Swindon 1959
51043	DMBSL	Bo'ness & Kinneil Railway	Swindon 1959
59404	TCK	Bo'ness & Kinneil Railway	Swindon 1959
79443	TRCsoKBL	Bo'ness & Kinneil Railway	Swindon 1957

59404 was converted by BR to TSK –/56 but restored as TCK.

CLASS 127 DERBY SUBURBAN 4-CAR UNITS

Built: 1959. Non-gangwayed. Normal formation: DMBS–TSL–TS–DMBS. Some DMBS rebuilt as DMPMV Normal formation: DMPMV(A)–DMPMV(B).
Engines: Two Rolls-Royce C8 of 177 kW (238 hp).
Transmission: Hydraulic.

DMBS	20.45 x 2.82 m	40.6 tonnes	–/76
DMPMV (A)	20.45 x 2.82 m	40 tonnes	
DMPMV (B)	20.45 x 2.82 m	40 tonnes	
TSL	20.45 x 2.82 m	30.5 tonnes	–/86

51592		DMBS	South Devon Railway	Derby 1959
51604		DMBS	South Devon Railway	Derby 1959
51616	"ALF BENNY"	DMBS	Great Central Railway	Derby 1959
51618		DMBS	Llangollen Railway	Derby 1959
51622		DMBS	Great Central Railway	Derby 1959

▲ Bo'ness & Kinneil Class 126 DMU 51043+59404+51017 has just left Kinneil with the 13.05 Bo'ness–Manuel on 29 August 2010. **Ian Lothian**

▼ Superbly restored Derby Lightweight single car 79900 stands at Wirksworth on 2 May 2011, ready to carry passengers up the 1 in 27 to Ravenstor. **Paul Abell**

51591–55966	DMPMV(A)	Midland Railway-Butterley	Derby 1959
51610–55967 "Glen Ord"	DMPMV(B)	Midland Railway-Butterley	Derby 1959
51625–55976	DMPMV(A)	Midland Railway-Butterley	Derby 1959
59603	TSL	Chasewater Light Railway	Derby 1959
59609	TSL	Midland Railway-Butterley	Derby 1959

UNCLASSIFIED DERBY LIGHTWEIGHT 2-CAR UNIT

Built: 1955. Normal formation. DMBS–DTCL.
Engines: Two BUT (AEC) of 112 kW (150 hp).

DMBS	18.49 x 2.82 m	27.4 tonnes	–/61
DTCL	18.49 x 2.82 m	21.3 tonnes	9/53

79018–DB 975007	DMBS	Midland Railway-Butterley	Derby 1955
79612–DB 975008	DTCL	Midland Railway-Butterley	Derby 1955

UNCLASSIFIED DERBY LIGHTWEIGHT SINGLE-CAR UNIT

Built: 1956. Non-gangwayed single cars with two driving cabs.
Engines: Two AEC of 112 kW (150 hp).

DMBS	18.49 x 2.82 m	27 tonnes	–/52

79900–DB975010 "IRIS"	DMBS	Ecclesbourne Valley Railway	Derby 1956

5.5. FOUR-WHEELED DIESEL RAILBUSES

UNCLASSIFIED WAGGON UND MASCHINENBAU

Built: 1958. 5 built.
Engine: Buessing of 112 kW (150 hp) at 1900 rpm (§ AEC 220 of 112 kW (150 hp)).
Maximum Speed: 70 mph.

DMS	13.95 x 2.67 m	15 tonnes	–/56

79960	North Norfolk Railway	WMD 1265/1958
79962	Keighley & Worth Valley Railway	WMD 1267/1958
79963	North Norfolk Railway	WMD 1268/1958
79964§	Keighley & Worth Valley Railway	WMD 1298/1958

UNCLASSIFIED AC CARS

Built: 1958. 5 built.
Engine: AEC 220 of 112 kW (150 hp) (§ engine removed).
Maximum Speed: 70 mph.

DMS	11.33 x 2.82 m	11 tonnes	–/46

79976§	Great Central Railway	AC 1958
79978	Colne Valley Railway	AC 1958

UNCLASSIFIED BR DERBY/LEYLAND

Built: 1977.
fitted 1979).
Transmission: Mechanical. Self Changing Gears.
Maximum Speed: 75 mph.

Air braked.

DMS	12.32 x 2.50 m	16.67 tonnes	–/40

R1-RDB 975874	North Norfolk Railway (N)	RTC Derby 1977

UNCLASSIFIED BREL DERBY/LEYLAND/WICKHAM

Built: 1980.
Engine: Leyland 690 of 149 kW (200 hp).
Transmission: Mechanical. Self Changing Gears.
Maximum Speed: 75 mph.

Air braked.

DMS	15.30 x 2.50 m	19.8 tonnes	–/56

R3.01	Connecticut Trolley Museum, East Windsor, CT, USA	Wkm 1980

UNCLASSIFIED BREL DERBY/LEYLAND

Built: 1981.
Engine: Leyland 690 of 149 kW (200 hp).
Transmission: Mechanical. Self Changing Gears SE4 epicyclic gearbox and cardan shafts to SCG RF28 final drive.
Maximum Speed: 75 mph.
Gauge: Built as 1435 mm but converted to 1600 mm when sold to Northern Ireland Railways.

Air braked.

DMS	15.30 x 2.50 m	19.96 tonnes	–/56

R3.03–RDB 977020	RB3	Downpatrick Steam Railway, NI	RTC Derby 1981

UNCLASSIFIED BREL-LEYLAND

Built: 1984.
Engine: Leyland TL11 of 152 kW (205 hp).
Transmission: Mechanical.
Maximum Speed: 75 mph.

Air braked.

DMS	x 2.50 m	37.5 tonnes	–/64 (§ –/40)

RB 002 "DENMARK"	Riverstown Mill Railway, Dundalk, Ireland	BREL-Leyland 1984
RB 004 § "USA"	Aln Valley Railway	BREL-Leyland 1984

CLASS 140 DERBY/LEYLAND BUS PROTOTYPE 2-CAR RAILBUS

Built: 1981.
Engine: Leyland TL11 of 152 kW (205 hp)
Transmission: Mechanical. Self-Changing Gears 4-speed gearbox.
Maximum Speed: 75 mph.

Air braked.

DMSL	16.20 x 2.50 m	23.2 tonnes	–/50
DMS	16.20 x 2.50 m	23.0 tonnes	–/52

55500	DMS	Keith & Dufftown Railway	Derby 1981
55501	DMSL	Keith & Dufftown Railway	Derby 1981

▲ Four-wheeled Diesel Railbus 79964 leaves Keighley heading for Oxenhope on 7 March 2010.
Brian Dobbs

▼ Leyland Experimental Vehicle 975874 (or LEV1) is preserved at the North Norfolk Railway. It is seen on display at Weybourne on 16 August 2009. **Jamie Squibbs**

CLASS 141 BREL/LEYLAND BUS 2-CAR RAILBUS

Built: 1983–84. Modified by Andrew Barclay 1988–89.
Engine: Leyland TL11 of 152 kW (205 hp).
Transmission: Hydraulic. Voith T211r.
Maximum Speed: 75 mph.

Air braked.

DMS	15.45 x 2.50 m	26.0 tonnes	–/50
DMSL	15.45 x 2.50 m	26.5 tonnes	–/44

55503	(ex-unit 141 103)	DMS	Weardale Railway	BREL-Leyland 1984
55508	(ex-unit 141 108)	DMS	Colne Valley Railway	BREL-Leyland 1984
55510	(ex-unit 141 110)	DMS	Weardale Railway	BREL-Leyland 1984
55513	(ex-unit 141 113)	DMS	Weardale Railway	BREL-Leyland 1984
55523	(ex-unit 141 103)	DMSL	Weardale Railway	BREL-Leyland 1984
55528	(ex-unit 141 108)	DMSL	Colne Valley Railway	BREL-Leyland 1984
55533	(ex-unit 141 113)	DMSL	Weardale Railway	BREL-Leyland 1984

In addition a further 28 Class 141 vehicles have been exported for use abroad:

The following vehicles have been sold to Iranian Islamic Republic Railways: 55502, 55505, 55507, 55509, 55511, 55514, 55515, 55516, 55517, 55518, 55519, 55520, 55522, 55525, 55527, 55529, 55531, 55534, 55535, 55536, 55537, 55538, 55539 and 55540.

The following vehicles have been sold to Connexion, Utrecht, Netherlands: 55506, 55512, 55526 and 55532.

UNCLASSIFIED WICKHAM TRACK RECORDING CAR

Built: 1958 for BR Research. Later known as the Wickham Self-propelled Laboratory. Preserved as passenger-carrying Railbus.
Engine: **Transmission:** Mechanical.
Maximum Speed:

999507-RDB 999507	Lavender Line, Isfield	Wkm 1958

▶ In County Durham the Weardale Railway normally uses Pacers for its community services. 141 103 (55503+55523) stands at Stanhope with a train for Bishop Auckland on 29 January 2011. **Paul Abell**

5.6. DIESEL ELECTRIC MULTIPLE UNITS

For old set numbers, the last set that that vehicle was formed in is given.

CLASS 201 "HASTINGS" 6-CAR DIESEL-ELECTRIC UNITS

Built: 1957 by BR Eastleigh Works on frames constructed at Ashford. Special narrow-bodied units built to the former loading gauge of the Tonbridge–Hastings line.
Normal formation: DMBSO–TSOL–TSOL–TFK–TSOL–DMBSO.
Engines: English Electric 4SRKT of 370 kW (500 hp).
Transmission: Two EE 507 traction motors on the inner power car bogie.
Maximum Speed: 75 mph.

DMBSO	18.35 x 2.50 m	54 tonnes	–/22
TSOL	18.35 x 2.50 m	29 tonnes	–/52
TFK	18.36 x 2.50 m	30 tonnes	42/–

60000	"Hastings"	DMBSO	St Leonards Railway Engineering	Eastleigh 1957
60001		DMBSO	St Leonards Railway Engineering	Eastleigh 1957
60500		TSOL	St Leonards Railway Engineering	Eastleigh 1957
60501		TSOL	St Leonards Railway Engineering	Eastleigh 1957
60502		TSOL	St Leonards Railway Engineering	Eastleigh 1957
60700		TFK	St Leonards Railway Engineering	Eastleigh 1957

CLASS 202 "HASTINGS" 6-CAR DIESEL-ELECTRIC UNITS

Built: 1957–58 by BR Eastleigh Works on frames constructed at Ashford. Special narrow-bodied units built to the former loading gauge of the Tonbridge–Hastings line.
Normal formation: DMBSO–TSOL–TSOL (or TRSKB)–TFK–TSOL–DMBSO.
Engines: English Electric 4SRKT of 370 kW (500 hp).
Transmission: Two EE 507 traction motors on the inner power car bogie.
Maximum Speed: 75 mph.

DMBSO	20.34 x 2.50 m	55 tonnes	–/30
TSOL	20.34 x 2.50 m	29 tonnes	–/60
TFK	20.34 x 2.50 m	31 tonnes	48/–
TRSKB	20.34 x 2.50 m	34 tonnes	–/21

60016	"60116 Mountfield"	DMBSO	St Leonards Railway Engineering	Eastleigh 1957
60018	"60118 Tunbridge Wells"	DMBSO	St Leonards Railway Engineering	Eastleigh 1957
60019		DMBSO	St Leonards Railway Engineering	Eastleigh 1957
60527		TSOL	St Leonards Railway Engineering	Eastleigh 1957
60528		TSOL	St Leonards Railway Engineering	Eastleigh 1957
60529		TSOL	St Leonards Railway Engineering	Eastleigh 1957
60708		TFK	St Leonards Railway Engineering	Eastleigh 1957
60709		TFK	St Leonards Railway Engineering	Eastleigh 1957
60750–RDB 975386		TRSKB	The Pump House Steam & Transport Museum	Eastleigh 1958

CLASS 205 "HAMPSHIRE" 3-CAR DIESEL-ELECTRIC UNITS

Built: 1958–59 by BR Eastleigh Works on frames constructed at Ashford. Non-gangwayed.
Normal formation: DMBSO–TSO–DTCsoL.
Engines: English Electric 4SRKT of 370 kW (500 hp).
Transmission: Two EE 507 traction motors on the inner power car bogie.
Maximum Speed: 75 mph.

DMBSO	20.34 x 2.82 m	56 tonnes	–/52
TSO	20.28 x 2.82 m	30 tonnes	–/104
DTCsoL	20.34 x 2.82 m	32 tonnes	19/50

60100–60154	DMBSO	(ex-unit 001)	East Kent Light Railway	Eastleigh 1957
60108	DMBSO	(ex-unit 009)	Eden Valley Railway	Eastleigh 1957
60110	DMBSO	(ex-unit 205)	Epping–Ongar Railway	Eastleigh 1957
60117	DMBSO	(ex-unit 018)	Pontypool & Blaenavon Railway	Eastleigh 1957

60122	DMBSO	(ex-unit 023)	Lavender Line	Eastleigh 1959
60124	DMBSO	(ex-unit 025)	Mid Hants Railway	Eastleigh 1959
60145–977939	DMBSO	(ex-unit 027)	St Leonards Railway Engineering	Eastleigh 1962
60146	DMBSO	(ex-unit 028)	Dartmoor Railway	Eastleigh 1962
60149–977940	DMBSO	(ex-unit 031)	St Leonards Railway Engineering	Eastleigh 1962
60150	DMBSO	(ex-unit 032)	Dartmoor Railway	Eastleigh 1962
60151	DMBSO	(ex-unit 033)	Lavender Line	Eastleigh 1962
60658	TSO	(ex-unit 009)	Eden Valley Railway	Eastleigh 1959
60669	TSO	(ex-unit 024)	Marshalls Transport, Pershore Airfield	Eastleigh 1959
60673	TSO	(ex-unit 028)	Dartmoor Railway	Eastleigh 1962
60677	TSO	(ex-unit 032)	Dartmoor Railway	Eastleigh 1962
60678	TSO	(ex-unit 033)	Cold Norton Play School, Essex	Eastleigh 1962
60800	DTCsoL	(ex-unit 001)	East Kent Light Railway	Eastleigh 1957
60808	DTCsoL	(ex-unit 009)	Eden Valley Railway	Eastleigh 1957
60810	DTCsoL	(ex-unit 205)	Epping–Ongar Railway	Eastleigh 1957
60820	DTCsoL	(ex-unit 021)	Lavender Line	Eastleigh 1958
60822	DTCsoL	(ex-unit 023)	Marshalls Transport, Pershore Airfield	Eastleigh 1959
60824	DTCsoL	(ex-unit 025)	Mid Hants Railway	Eastleigh 1959
60827	DTCsoL	(ex-unit 028)	Dartmoor Railway	Eastleigh 1962
60828	DTCsoL	(ex-unit 018)	Pontypool & Blaenavon Railway	Eastleigh 1962
60831	DTCsoL	(ex-unit 032)	Dartmoor Railway	Eastleigh 1962
60832	DTCsoL	(ex-unit 033)	Lavender Line	Eastleigh 1962

CLASS 207 "OXTED" 3-CAR DIESEL-ELECTRIC UNITS

Built: 1962 by BR Eastleigh Works on frames constructed at Ashford. Reduced body width to allow operation through Somerhill Tunnel. Non-gangwayed.
Normal formation: DMBSO–TCsoL–DTSO.
Engines: English Electric 4SRKT of 370 kW (500 hp).
Transmission: Two EE 507 traction motors on the inner power car bogie.
Maximum Speed: 75 mph.

DMBSO	20.34 x 2.74 m	56 tonnes	–/42
DTSO	20.32 x 2.74 m	32 tonnes	–/76
TCsoL	20.34 x 2.74 m	31 tonnes	24/42

60127	DMBSO	(ex-unit 203)	Swindon & Cricklade Railway	Eastleigh 1962
60130	DMBSO	(ex-unit 202)	East Lancashire Railway	Eastleigh 1962
60138–977907	DMBSO	(ex-unit 013)	The Pump House Steam & Transport Museum	Eastleigh 1962
60142	DMBSO	(ex-unit 017)	Spa Valley Railway	Eastleigh 1962
60616	TCsoL	(ex-unit 017)	Spa Valley Railway	Eastleigh 1962
60901	DTSO	(ex-unit 201)	Swindon & Cricklade Railway	Eastleigh 1962
60904	DTSO	(ex-unit 202)	East Lancashire Railway	Eastleigh 1962
60916	DTSO	(ex-unit 017)	Spa Valley Railway	Eastleigh 1962

CLASS 210 DERBY PROTOTYPE 4-CAR DIESEL-ELECTRIC UNITS

Built: 1981 by BR Derby Works.
Normal formation: DMBSO–TSO–TSOL–DTSO.
Engines: MTU 12V396TC11 of 850 kW.
Transmission:
Maximum Speed: 75 mph.

DTSO	20.52 x 2.82 m	29 tonnes	–/74

54000–60300–67300	DTSO	Electric Railway Museum, Coventry	Derby 1981

6. ELECTRIC MULTIPLE UNITS

TYPE CODES

For type codes see section 5.

6.1. SOUTHERN RAILWAY EMU STOCK

CLASS 487 WATERLOO & CITY LINE UNITS

Built: 1940. No permanent formations.
System: 630 V DC third rail.
Traction Motors: Two EE 500 of 140 kW (185 hp). **Maximum Speed:** 35 mph.

DMBTO	14.33 x 2.64 m	29 tons	–/40

BR	*SR*			
S61S	61	DMBTO	London Transport Depot Museum, Acton (N)	EE 1940

1285 CLASS (later 3 Sub) SUBURBAN UNITS

Built: 1925. Normal formation: DMBT–TT–DMBT.
System: 630 V DC third rail.
Traction Motors: Two MV 167 kW (225 hp). **Maximum Speed:** 75 mph.

DMBT	18.90 x 2.44 m	39 tons	–/70

BR	*SR*				
S8143S	8143	DMBT	(ex-unit 1293 later 4308)	National Railway Museum, York (N)	MC 1925

4 Cor "NELSONS" PORTSMOUTH EXPRESS STOCK

Built: 1937–38. Normal formation: DMBTO–TTK–TCK–DMBTO.
System: 630 V DC third rail.
Traction Motors: Two MV 167 kW (225 hp) per power car.
Maximum Speed: 75 mph.

DMBTO	19.54 x 2.88 m	46.5 tons	–/52
TTK	19.54 x 2.85 m	32.65 tons	–/68
TCK	19.54 x 2.85 m	32.6 tons	30/24

BR	*SR*				
S10096S	10096	TTK	(ex-unit 3142)	East Kent Light Railway	Eastleigh 1937
S11161S	11161	DMBTO	(ex-unit 3142)	East Kent Light Railway	Eastleigh 1937
S11179S	11179	DMBTO	(ex-unit 3131)	National Railway Museum, York (N)	Eastleigh 1937
S11187S	11187	DMBTO	(ex-unit 3135)	London Transport Depot Museum, Acton	Eastleigh 1937
S11201S	11201	DMBTO	(ex-unit 3142)	Bluebell Railway	Eastleigh 1937
S11825S	11825	TCK	(ex-unit 3142)	East Kent Light Railway	Eastleigh 1937

S11161S was originally in unit 3065 and S11825S in unit 3135.

4 Sub (later Class 405) — SUBURBAN UNITS

Built: 1941–51. Normal formation: DMBTO–TT–TTO–DMBTO.
System: 630 V DC third rail.
Traction Motors: Two EE507 of 185 kW (250 hp). **Maximum Speed:** 75 mph.

DMBTO	19.05 x 2.82 m	42 tons	–/82
TT	18.90 x 2.82 m	27 tons	–/120
TTO	18.90 x 2.82 m	26 tons	–/102

BR	*SR*				
S 10239 S	10239	TT	(ex-unit 4732)	Electric Railway Museum, Coventry	Eastleigh 1947
S 12354 S		TTO	(ex-unit 4732)	Electric Railway Museum, Coventry	Eastleigh 1948
S 12795 S		DMBTO	(ex-unit 4732)	Electric Railway Museum, Coventry	Eastleigh 1951
S 12796 S		DMBTO	(ex-unit 4732)	Electric Railway Museum, Coventry	Eastleigh 1951

S 10239 S was originally in unit 4413 and S 12354 S in unit 4381.

2 Bil — SEMI-FAST UNITS

Built: 1937. Normal formation: DMBTK–DTCK.
System: 630 V DC third rail.
Traction Motors: Two EE of 205 kW (275 hp). **Maximum Speed:** 75 mph.

DMBTK	19.24 x 2.85 m	43.5 tons	–/52
DTCK	19.24 x 2.85 m	31.25 tons	24/30

BR	*SR*				
S 10656 S	10656	DMBTK	(ex-unit 1890–2090)	National Railway Museum, Shildon (N)	Eastleigh 1937
S 12123 S	12123	DTCK	(ex-unit 1890–2090)	National Railway Museum, Shildon (N)	Eastleigh 1937

4 DD — DOUBLE-DECK SUBURBAN UNITS

Built: 1949. Normal formation: DMBT–TT–TT–DMBT.
System: 630 V DC third rail.
Traction Motors: Two EE of 185 kW (250 hp). **Maximum Speed:** 75 mph.

DMBT	19.24 x 2.85 m	39 tons	–/121

S 13003 S	DMBT	(ex-unit 4002–4902)	Hope Farm, Sellindge	Lancing 1949
S 13004 S	DMBT	(ex-unit 4002–4902)	Northamptonshire Ironstone Railway	Lancing 1949

4 EPB (later Class 415) — SUBURBAN UNITS

Built: 1951–57. Normal formation: DMBTO–TT–TTO–DMBTO.
System: 630 V DC third rail.
Traction Motors: Two EE507 of 185 kW (250 hp). **Maximum Speed:** 75 mph.

DMBTO	19.05 x 2.82 m	42 tons	–/82
TTO	18.90 x 2.82 m	26 tons	–/102

S 14351 S	DMBTO	(ex-unit 5176)	Northamptonshire Ironstone Railway	Eastleigh 1955
S 15354 S	TTO	(ex-unit 5176)	Electric Railway Museum, Coventry	Eastleigh 1955
S 15396 S	TTO	(ex-unit 5176)	Northamptonshire Ironstone Railway	Eastleigh 1956
S 14352 S	DMBTO	(ex-unit 5176)	Northamptonshire Ironstone Railway	Eastleigh 1955

S 15396 S was originally in unit 5208.

2 EPB (later Class 416/1) — SUBURBAN UNITS

Built: 1959. Normal formation: DMBSO–DTSO.
System: 630 V DC third rail.
Traction Motors: Two EE of 185 kW (250 hp). **Maximum Speed:** 75 mph.

DMBSO	19.05 x 2.82 m	40 tons	–/82
DTSO	18.90 x 2.82 m	30 tons	–/92

S 14573 S	DMBSO	(ex-unit 5667–6307)	Electric Railway Museum, Coventry	Eastleigh 1959
S 16117 S	DTSO	(ex-unit 5667–6307)	Electric Railway Museum, Coventry	Eastleigh 1959

6.2. PULLMAN CAR COMPANY EMU STOCK

GENERAL

Pullman cars owned by the Pullman Car Company operated as parts of EMU formations on the Southern Railway (later BR Southern Region). In addition the three Brighton Belle EMU sets were composed entirely of Pullman vehicles.

All vehicles are used as hauled stock except * – static exhibits.

6 Pul

Built: 1932. 6-car sets incorporating one Pullman kitchen composite. Normal formation: DMBTO–TTK–TCK–TPCK–TCK–DMBTO.

TPCK	20.40 x 2.77 m	43 tons	12/16

RUTH	S 264 S	TPCK	(ex-unit 2017–3042)	VSOE, Stewarts Lane, London	MC 1932
BERTHA	S 278 S	TPCK	(ex-unit 2012–3001)	West Coast Railway Co., Carnforth	MC 1932

5 Bel — BRIGHTON BELLE UNITS

Built: 1932. 5-car all Pullman sets. Formation: DMPBT–TPT–TPKF–TPKF–DMPBT.
System: 630 V DC Third rail.
Traction Motors: Four BTH of 167 kW (225 hp).

TPFK	20.40 x 2.77 m	42 tons	20/–
TPT	20.40 x 2.77 m	41 tons	–/56
DMPBT	20.62 x 2.77 m	62 tons	–/48

HAZEL*	S279S	TPFK	(ex-unit 2051–3051)	Black Bull Inn, Moulton, N Yorks	MC 1932
AUDREY	S280S	TPFK	(ex-unit 2052–3052)	VSOE, Stewarts Lane, London	MC 1932
GWEN	S281S	TPFK	(ex-unit 2053–3053)	VSOE, Stewarts Lane, London	MC 1932
DORIS	S282S	TPFK	(ex-unit 2051–3051)	Bluebell Railway	MC 1932
MONA	S283S	TPFK	(ex-unit 2053–3053)	VSOE, Stewarts Lane, London	MC 1932
VERA	S284S	TPFK	(ex-unit 2052–3052)	VSOE, Stewarts Lane, London	MC 1932
CAR No. 85	S285S	TPT	(ex-unit 2053–3053)	Rampart C&W Services, Derby	MC 1932
CAR No. 86	S286S	TPT	(ex-unit 2051–3051)	VSOE, Stewarts Lane, London	MC 1932
CAR No. 87	S287S	TPT	(ex-unit 2052–3052)	Southall Depot, London	MC 1932
CAR No. 88	S288S	DMPBT	(ex-unit 2051–3051)	Barrow Hill Roundhouse	MC 1932
CAR No. 89*	S289S	DMPBT	(ex-unit 2051–3051)	Little Mill Inn, Rowarth, Derbys.	MC 1932
CAR No. 91	S291S	DMPBT	(ex-unit 2052–3052)	Barrow Hill Roundhouse	MC 1932
CAR No. 92	S292S	DMPBT	(ex-unit 2053–3053)	VSOE, Stewarts Lane, London	MC 1932
CAR No. 93	S293S	DMPBT	(ex-unit 2053–3053)	VSOE, Stewarts Lane, London	MC 1932

The following vehicles are undergoing restoration for the Brighton Belle project: Nos. S282S, S285S, S287S, S288S, S291S.

▶ In one of the least likely locations which appear in this book, Brighton Belle Car No. 89 can be seen outside the Little Mill Inn, Rowarth, Derbyshire, as on 7 November 2010. (The pub meals can be recommended.) **Paul Abell**

6.3. LMS & CONSTITUENTS EMU STOCK

LNWR EUSTON–WATFORD STOCK

Built: 1915. Oerlikon design. Normal formation: DMBTO–TTO–DTTO.
System: 630 V DC third rail. Used on Euston–Watford line.
Traction Motors: Four Oerlikon 179 kW (240 hp).
Maximum Speed:

DMBTO 17.60 x 2.73 m 54.75 tonnes –/48

BR	*LMS*			
M 28249 M	28249	DMBTO	National Railway Museum, York (N)	MC 1915

CLASS 502 LIVERPOOL–SOUTHPORT STOCK

Built: 1939. Normal formation: DMBTO–TTO–DTTO (originally DTCO).
System: 630 V DC third rail.
Traction Motors: Four EE 175 kW. **Maximum Speed:** 65 mph.

DMBTO	21.18 x 2.90 m	42.5 tonnes	–/88
DTTO	21.18 x 2.90 m	25.5 tonnes	–/79 (built as DTCO 53/25)

BR	*LMS*			
M 28361 M	28361	DMBTO	Tebay, Cumbria	Derby 1939
M 29896 M	29896	DTTO	Tebay, Cumbria	Derby 1939

CLASS 503 MERSEY WIRRAL STOCK

Built: 1938. Normal formation: DMBTO–TTO (originally TCO)–DTTO.
System: 630 V DC third rail.
Traction Motors: 4 BTH 100 kW. **Maximum Speed:** 65 mph.

DMBTO	18.48 x 2.77 m	36.5 tonnes	–/56
TTO	17.77 x 2.77 m	20.5 tonnes	–/58 (built as TCO 40/19)
DTTO	18.85 x 2.77 m	21.5 tonnes	–/66

BR	*LMS*			
M 28690 M	28690	DMBTO	Electric Railway Museum, Coventry	Derby 1938
M 29720 M	29720	TTO	Electric Railway Museum, Coventry	Derby 1938
M 29289 M	29289	DTTO	Electric Railway Museum, Coventry	Derby 1938

MSJ&A STOCK

Built: 1931. Normal formation: DMBT–TC–DTT.
System: 1500 V DC overhead. Used on Manchester South Junction and Altrincham line until it was converted to 25 kV AC. This line is now part of the Manchester Metrolink system.
Traction Motors: **Maximum Speed:** 65 mph.

TC 17.60 x 2.85 m 31 tonnes. 24/72

BR	*LMS*	*MSJ&A*			
M 29666 M	29666	117	TC	Midland Railway-Butterley	MC 1931
M 29670 M	29670	121	TC	Midland Railway-Butterley	MC 1931

6.4. LNER & CONSTITUENTS EMU STOCK

NORTH EASTERN RAILWAY DMLV

Built: 1904. Driving motor luggage van for North Tyneside line. After withdrawal from capital stock, this vehicle was used as a rail de-icing car.
System: 675 V DC third rail.
Traction Motors: **Maximum Speed**:

DMLV 17.40 x 2.77 m 46.5 tonnes

BR	*LNER*	*NER*			
DE 900730	23267	3267	DMLV	Stephenson Railway Museum (N)	MC 1904

GRIMSBY & IMMINGHAM LIGHT RAILWAY TRAM A1-1A

Built: 1915 by GCR Dukinfield.
Type: Single deck tram.
Seats: 64 + 8 tip-up.
Bogies: Brush.
Motors: 2 x 25 h.p. Dick Kerr DK9 of 18 kW.

14 Crich Tramway Village GCR Dukinfield 1915

CLASS 306 LIVERPOOL STREET–SHENFIELD STOCK

Built: 1949. Normal formation: DMSO–TBSO–DTSO.
System: 25 kV AC overhead (originally 1500 V DC overhead).
Traction Motors: Four Crompton Parkinson of 155 kW.
Maximum Speed: 65 mph.

DMSO	18.41 x 2.90 m	51.7 tonnes	–/62
TBSO	16.78 x 2.90 m	26.4 tonnes	–/46
DTSO	16.87 x 2.90 m	27.9 tonnes	–/60

E 65217 E	DMSO	Knights Rail Services, Eastleigh Works	MC 1949
E 65417 E	TBSO	Knights Rail Services, Eastleigh Works	MC 1949
E 65617 E	DTSO	Knights Rail Services, Eastleigh Works	BRCW 1949

▶ Grimsby & Immingham Light Railway tram 14 was on display at the Crich Tramway Museum Enthusiasts' Day on 29 May 2011, with Gateshead tram 5 (which ran on the Grimsby & Immingham as No. 20) just visible alongside. **Paul Abell**

6.5. BRITISH RAILWAYS EMU STOCK

BR EMU power cars were numbered in the 6xxxx series starting with 61000 whilst trailer cars were numbered in the 7xxxx series. This does not apply to the APT-P or the battery EMU.

CLASS 302 — BR

Built: 1958–60 for London Fenchurch Street–Shoeburyness services.
System: 25 kV AC overhead.
Original Formation: BDTSOL–MBS–TCsoL–DTS.
Formation as rebuilt: BDTCOL–MBSO–TSOL–DTSO.
Traction Motors: Four English Electric EE 536A of 143.5 kW.
Maximum Speed: 75 mph.

DTSO	20.36 x 2.83 m	33.4 tonnes	–/88

75033	DTSO	(ex-unit 201)	Mangapps Railway Museum	York 1958
75250	DTSO	(ex-unit 277)	Mangapps Railway Museum	York 1959

CLASS 303 — PRESSED STEEL

Built: 1959–61 for Glasgow area services.
System: 25 kV AC overhead.
Formation: DTSO–MBSO–BDTSO.
Traction Motors: Four MV of 155 kW.
Maximum Speed: 75 mph.

DTSO	20.18 x 2.83 m	34.4 tonnes	–/83
MBSO	20.18 x 2.83 m	56.4 tonnes	–/70
BDTSO	20.18 x 2.83 m	38.4 tonnes	–/83

61503	MBSO	(ex-unit 023)	Bo'ness & Kinneil Railway	PS 1960
75597	DTSO	(ex-unit 023)	Bo'ness & Kinneil Railway	PS 1960
75632	BDTSO	(ex-unit 032)	Bo'ness & Kinneil Railway	PS 1960

CLASS 307 — BR

Built: 1954–56 for London Liverpool Street–Southend Victoria services.
System: 1500 V DC overhead. Converted 1960–61 to 25 kV AC overhead.
Original Formation: BDTBS–MS–TCsoL–DTSOL.
Formation as rebuilt: BDTBSO–MSO–TSOL–DTCOL.
Traction Motors: Four GEC WT344 of 130 kW.
Maximum Speed: 75 mph.

BDTBSO	20.18 x 2.83 m	43 tonnes	–/66
DTCO	20.18 x 2.83 m	33 tonnes	24/48

75023	BDTBSO	(ex-unit 123)	Electric Railway Museum, Coventry	Eastleigh 1956
75120–94320	DTCO	(ex-unit 120)	Mid Norfolk Railway	Eastleigh 1956

75120 was rebuilt as propelling control vehicle.

CLASS 308 — BR

Built: 1961 for London Liverpool Street–Clacton stopping services.
System: 25 kV AC overhead.
Original Formation: BDTCOL–MBS–TCSoL–DTS.
Formation as rebuilt: BDTCOL–MBSO–TSOL–DTSO.
Traction Motors: Four English Electric EE 536A of 143.5 kW.
Maximum Speed: 75 mph.

BDTCOL	20.18 x 2.82 m	36.3 tonnes	24/52

75881	BDTCOL	(ex-unit 136)	The Pump House Steam & Transport Museum	York 1961

▲ Built by the LNWR for Euston–Watford services in 1915, DMBTO M28249 is seen in the Great Hall of NRM York on 22 January 2011. **Robert Pritchard**

▼ Two Class 302 EMU vehicles were preserved and are based at the Mangapps Railway Museum, where they are used as hauled stock. On 29 August 2009 75033+75250 are propelled by a Class 03. 33202 and 47793 are to the right. **Phil Barnes**

CLASS 309 — BREL YORK

Built: 1962–63 for London Liverpool Street–Clacton express services.
System: 25 kV AC overhead.
Original formation: BDTCSOL–MBSOL–TSO–DTCSO.
Formation as rebuilt: BDTCsoL–MBSOL–TSO–DTSOL.
Traction Motors: Four GEC WT401 of 210 kW.
Maximum Speed: 100 mph.

MBSOL	20.18 x 2.82 m	57.7 tonnes	–/52
BDTCsoL	20.18 x 2.82 m	40.0 tonnes	18/32
DTSOL	20.18 x 2.82 m	36.6 tonnes	–/56

61928–977966	MBSOL	(ex-unit 624)	Electric Railway Museum, Coventry	York 1962
61937–977963	MBSOL	(ex-unit 616)	Electric Railway Museum, Coventry	York 1962
75642–977962	BDTCsoL	(ex-unit 616)	Electric Railway Museum, Coventry	York 1962
75965–977965	BDTCsoL	(ex-unit 624)	Electric Railway Museum, Coventry	York 1962
75972–977967	DTSOL	(ex-unit 624)	Electric Railway Museum, Coventry	York 1962
75981–977964	DTSOL	(ex-unit 616)	Electric Railway Museum, Coventry	York 1963

61928 is named NEW DALBY.

CLASS 311 — CRAVENS

Built: 1967 for Glasgow "South Side electrification" extension to Gourock and Wemyss Bay.
System: 25 kV AC overhead.
Formation: DTSO(A)–MBSO–DTSO(B).
Traction Motors: Four AEI of 165 kW. **Maximum Speed:** 75 mph.

MBSO	20.18 x 2.83 m	56.4 tonnes	–/70
DTSO(A)	20.18 x 2.83 m	34.4 tonnes	–/83

62174–977845	MBSO	(ex-unit 103)	Summerlee Museum of Scottish Industrial Life	Cravens 1967
76414–977844	DTSO(A)	(ex-unit 103)	Summerlee Museum of Scottish Industrial Life	Cravens 1967

CLASS 312 — BREL YORK

Built: 1975–78 using Mark 2 bodyshell for outer-suburban services from London King's Cross and London Liverpool Street and in the West Midlands area.
System: 25 kV AC overhead.
Formation: BDTSO–MBSO–TSO–DTCO.
Traction Motors: Four English Electric 546 of 201.5 kW.
Maximum Speed: 90 mph.

TSO	20.18 x 2.82 m	30.5 tonnes	–/98
DTCO	20.18 x 2.82 m	33.0 tonnes	25/47

71205	TSO	(ex-unit 792)	Electric Railway Museum, Coventry	York 1976
78037	DTCO	(ex-unit 792)	Electric Railway Museum, Coventry	York 1976

CLASS 370 PROTOTYPE ADVANCED PASSENGER TRAIN (APT-P)

Built: 1978–80. Designed to run as pairs of six-car articulated units with two power cars in the middle, these electric trains featured active hydraulic tilt and proved to be a maintenance nightmare. The power cars were reasonably successful, and are partly the basis of the Class 91 electric locomotive.
System: 25 kV AC overhead.
Normal Formation of Trailer Rake: DTSOL–TSOL–TSRB–TUOL–TFOL–TBFOL.
Formation of preserved set: DTSOL–TBFOL–M–TRSB–TBFOL–DTSOL.
Traction Motors: Four ASEA LJMA 410F body mounted.
Wheel Dia: 853 mm.
Maximum Speed: 125 mph.

DTSOL	21.44 x 2.72 m	33.7 tonnes	–/52
TBFOL	21.20 x 2.72 m	31.9 tonnes	25/–
TRSBL	21.20 x 2.72 m	26.75 tonnes	–/28
M	20.40 x 2.72 m	67.5 tonnes	

48103	DTSOL	Crewe Heritage Centre	Derby 1978
48106	DTSOL	Crewe Heritage Centre	Derby 1979
48602	TBFOL	Crewe Heritage Centre	Derby 1978
48603	TBFOL	Crewe Heritage Centre	Derby 1978
48404	TSRBL	Crewe Heritage Centre	Derby 1979
49002	M	Crewe Heritage Centre	Derby 1979
49006	M	National Railway Museum, Shildon (N)	Derby 1980

CLASSES 410 & 411 (4 Bep & 4 Cep) BR

Built: 1956–63 for Kent Coast electrification. Rebuilt 1979–84. Class 410 (4Bep) was later reclassified Class 412.
System: 750 V DC third rail.
Original Formation: DMBSO–TCK–TSK (4Cep) TRSB (4Bep)–DMBSO.
Formation as Rebuilt: DMSO–TBCK–TSOL (4Cep) TRSB (4Bep)–DMSO.
Traction Motors: Two English Electric EE507 of 185 kW.
Maximum Speed: 90 mph.

DMSO	20.34 x 2.82 m	49 tonnes	–/56
TSOL	20.18 x 2.82 m	36 tonnes	–/64
TBCK	20.18 x 2.82 m	34 tonnes	24/8
TRSB	20.18 x 2.82 m	35.5 tonnes	–/24

61229	DMSO	(ex-units 7105–1537)	East Kent Light Railway	Eastleigh 1958
61230	DMSO	(ex-units 7105–1537)	East Kent Light Railway	Eastleigh 1958
61736	DMSO	(ex-units 7175–2304–1198)	Pontypool & Blaenavon Railway	Eastleigh 1960
61737	DMSO	(ex-units 7175–2304–1198)	Pontypool & Blaenavon Railway	Eastleigh 1960
61742	DMSO	(ex-units 7178–1589)	Dartmoor Railway	Eastleigh 1960
61743	DMSO	(ex-units 7178–1589–1399)	Dartmoor Railway	Eastleigh 1960
61798	DMSO	(ex-units 7016–2305)	Eden Valley Railway	Eastleigh 1961
61799	DMSO	(ex-units 7016–2305)	Eden Valley Railway	Eastleigh 1961
61804	DMSO	(ex-units 7019–2301)	Eden Valley Railway	Eastleigh 1961
61805	DMSO	(ex-units 7019–2301)	Eden Valley Railway	Eastleigh 1961
69013	TRSB	(ex-units 7014–2305)	East Kent Light Railway	Eastleigh 1961
70229	TSOL	(ex-units 7105–1537)	Eden Valley Railway	Eastleigh 1958
70235	TBCK	(ex-units 7105–1537)	East Kent Light Railway	Eastleigh 1958
70262	TSOL	(ex-units 7113–1524)	St Leonards Railway Engineering	Eastleigh 1958
70273	TSOL	(ex-units 7124–1530)	Dartmoor Railway	Eastleigh 1958
70284	TSOL	(ex-units 7135–1520)	Northamptonshire Ironstone Railway	Eastleigh 1959
70292	TSOL	(ex-units 7143–1554)	St Modwen Properties, Long Marston	Eastleigh 1959
70294	TSOL	(ex-units 7145–1552)	HM Prison Lindholme, near Doncaster	Eastleigh 1959
70296	TSOL	(ex-units 7147–1559)	Northamptonshire Ironstone Railway	Eastleigh 1959
70300	TSOL	(ex-units 7151–1540)	Fighting Cocks PH, Middleton St George	Eastleigh 1959
70345	TBCK	(ex-units 7153–1500)	Hydraulic House, Sutton Bridge	Eastleigh 1959
70354	TBCK	(ex-units 7011–2305)	Eden Valley Railway	Eastleigh 1959
70508	TSOL	(ex-units 7159–1595–1399)	Dartmoor Railway	Eastleigh 1960
70510	TSOL	(ex-units 7161–1597)	Northamptonshire Ironstone Railway	Eastleigh 1960
70512	TSOL	(ex-units 7163–1605)	HM Prison Lindholme, near Doncaster	Eastleigh 1960
70527	TSOL	(ex-units 7178–1589)	Whitwell & Reepham Station, Norfolk	Eastleigh 1960
70531	TSOL	(ex-units 7152–1610)	St Modwen Properties, Long Marston	Eastleigh 1961
70539	TSOL	(ex-units 7190–1568)	Eden Valley Railway	Eastleigh 1961
70547	TSOL	(ex-units 7198–1569)	Garden Art, Bath Road, Hungerford	Eastleigh 1961
70549	TSOL	(ex-units 7200–1567)	East Lancashire Railway	Eastleigh 1961
70573	TBCK	(ex-units 7175–2304)	Pontypool & Blaenavon Railway	Eastleigh 1960
70576	TBCK	(ex-units 7178–1589)	Snibston Discovery Park, Leicestershire	Eastleigh 1960
70607	TBCK	(ex-units 7019–2301)	Eden Valley Railway	Eastleigh 1961

69013 also carried the number 69345 after rebuild as a TRSB.

70547 was latterly formed in DEMU 207 202 and 70549 was latterly formed in DEMU 207 203.

CLASS 414 (2 Hap) BR

Built: 1959 for the South Eastern Division of the former BR Southern Region.
System: 750 V DC third rail.
Formation: DMBSO–DTCsoL.
Traction Motors: Two English Electric EE507 of 185 kW.
Maximum Speed: 90 mph.

DMBSO	20.44 x 2.82 m	42 tonnes	–/84
DTCsoL	20.44 x 2.82 m	32.5 tonnes	19/60

61275	DMBSO	(ex-units 6077–4308)	National Railway Museum, York (N)	Eastleigh 1959
61287	DMBSO	(ex-units 6089–4311)	Electric Railway Museum, Coventry	Eastleigh 1959
75395	DTCsoL	(ex-units 6077–4308)	National Railway Museum, York (N)	Eastleigh 1959
75407	DTCsoL	(ex-units 6089–4311)	Electric Railway Museum, Coventry	Eastleigh 1959

CLASS 416/2 (2 EPB) BR

Built: 1954–56. 65373 & 77558 were built for the South Eastern Division of BR Southern Region. 65321 & 77112 were former North Eastern Region vehicles originally used between Newcastle and South Shields but were transferred to join the rest of the class on the Southern Region when the South Tyneside line was de-electrified in 1963.
System: 750 V DC third rail.
Formation: DMBSO–DTSso.
Traction Motors: Two English Electric EE507 of 185 kW.
Maximum Speed: 75 mph.

DMBSO	20.44 x 2.82 m	42 tonnes	–/82
DTSso	20.44 x 2.82 m	30.5 tonnes	–/102

65302–977874	DMBSO	(ex-units 6203–930 204)	Finmere Station, Oxfordshire	Eastleigh 1954
65304–977875	DMBSO	(ex-units 6205–930 204)	Finmere Station, Oxfordshire	Eastleigh 1954
65321–977505	DMBSO	(ex-units 5791–6291)	Electric Railway Museum, Coventry	Eastleigh 1955
65373	DMBSO	(ex-units 5759–6259)	East Kent Light Railway	Eastleigh 1956
65379–977925	DMBSO	(ex-units 6265–930 206)	Finmere Station, Oxfordshire	Eastleigh 1956
65382–977924	DMBSO	(ex-units 6268–930 206)	Finmere Station, Oxfordshire	Eastleigh 1956
77112–977508	DTSso	(ex-units 5793–6293)	Electric Railway Museum, Coventry	Eastleigh 1955
77558	DTSso	(ex-units 5759–6259)	East Kent Light Railway	Eastleigh 1956

CLASS 419 (MLV) BR

Built: 1959–61. Motor luggage vans for Kent Coast electrification. Fitted with traction batteries to allow operation on non-electrified lines.
System: 750 V DC third rail.
Traction Motors: Two English Electric EE507 of 185 kW.
Maximum Speed: 90 mph.

Also fitted with vacuum brakes for hauling parcels trains.

DMLV	20.45 x 2.82 m	45.5 tonnes

68001	(ex-units 9001–931 091)	East Kent Light Railway	Eastleigh 1959
68002	(ex-units 9002–931 092)	East Kent Light Railway	Eastleigh 1959
68003	(ex-units 9003–931 093)	Eden Valley Railway	Eastleigh 1961
68004	(ex-units 9004–931 094)	Mid Norfolk Railway	Eastleigh 1961
68005	(ex-units 9005–931 095)	Eden Valley Railway	Eastleigh 1961
68008	(ex-units 9008–931 098)	East Kent Light Railway	Eastleigh 1961
68009	(ex-units 9009–931 099)	East Kent Light Railway	Eastleigh 1961
68010	(ex-units 9010–931 090)	Wensleydale Railway	Eastleigh 1961

CLASS 420 & 421 (4 Big & 4 Cig) BR

Built: 1963–71 for the Central Division of BR Southern Region. Class 420 (4 Big) was later reclassified Class 422.
System: 750 V DC third rail.
Formation: DTCsoL–MBSO–TSO (4 Cig), TSRB (4 Big)–DTCsoL.
Traction Motors: Two English Electric EE507 of 185 kW.
Maximum Speed: 90 mph.

MBSO	20.18 x 2.82 m	49 tonnes	–/56
TSO	20.18 x 2.82 m	31.5 tonnes	–/72
DTCsoL	20.23 x 2.82 m	35 tonnes	24/28
TRSB	20.18 x 2.82 m	35 tonnes	–/40

62043	MBSO	(ex-units 7327–1127–1753)	The Moor Business Park, Ellough Moor	York 1965
62287	MBSO	(ex-units 7337–1237–1816–1303)	Lincolnshire Wolds Railway	York 1970
62364	MBSO	(ex-units 7376–1276–2251–1374)	Dean Forest Railway	York 1971
62378	MBSO	(ex-units 7390–1290–2255–1392)	Southall Depot, London	York 1971
62384	MBSO	(ex-units 7396–1296–2258–1353)	Great Central Railway	York 1971
62385	MBSO	(ex-units 7397–1297–2256–1399)	Pontypool & Blaenavon Railway	York 1971
62402	MBSO	(ex-units 7414–1214–1883–1497)	Mid Norfolk Railway	York 1971
62411	MBSO	(ex-units 7423–1223–1888–1498)	Swanage Railway	York 1972
69302	TRSB	(ex-units 7032–1276–2251)	Abbey View Disabled Day Centre, Neath	York 1963
69304	TRSB	(ex-units 7034–1299–2260)	Northamptonshire Ironstone Railway	York 1963
69306	TRSB	(ex-units 7036–1282–2254)	Spa Valley Railway	York 1963
69310	TRSB	(ex-units 7040–1290–2255)	Dartmoor Railway	York 1963
69316	TRSB	(ex-units 7046–1296–2258)	Waverley Route Heritage Association	York 1963
69318	TRSB	(ex-units 7048–1298–2259)	Colne Valley Railway	York 1963
69332	TRSB	(ex-units 7051–2203)	Dartmoor Railway	York 1969
69333	TRSB	(ex-units 7055–1802–2260)	Lavender Line	York 1969
69335	TRSB	(ex-units 7057–2209)	Wensleydale Railway	York 1969
69337	TRSB	(ex-units 7058–2210)	St Leonards Railway Engineering	York 1969
69338	TRSB	(ex-units 7054–2206)	Station Restaurant, Gulf Corporation, Bahrain	York 1969
69339	TRSB	(ex-units 7053–2205)	Great Central Railway	York 1969
70721	TSO	(ex-units 7327–1127–1753)	The Moor Business Park, Ellough Moor	York 1965
71041	TSO	(ex-units 7373–1273–1819–1306)	Hever Station, Kent	York 1971
71080	TSO	(ex-units 7412–1212–1881)	Dean Forest Railway	York 1971
71085	TSO	(ex-units 7417–1217–1884)	Deptford Station Cafe, London	York 1970
76048	DTCsoL	(ex-units 7327–1127–1753)	The Moor Business Park, Ellough Moor	York 1965
76102	DTCsoL	(ex-units 7327–1127–1753)	The Moor Business Park, Ellough Moor	York 1965
76726	DTCsoL	(ex-units 7376–1276–2251–1374)	Dean Forest Railway	York 1971
76740	DTCsoL	(ex-units 7390–1290–2255–1392)	Southall Depot, London	York 1971
76746	DTCsoL	(ex-units 7396–1296–2258–1393)	Great Central Railway	York 1971
76747	DTCsoL	(ex-units 7397–1297–2256–1399)	Pontypool & Blaenavon Railway	York 1971
76762	DTCsoL	(ex-units 7412–1212–1881)	Barrow Hill Roundhouse	York 1971
76764	DTCsoL	(ex-units 7414–1214–1883–1497)	Mid Norfolk Railway	York 1971
76773	DTCsoL	(ex-units 7423–1223–1888–1498)	Swanage Railway	York 1972
76797	DTCsoL	(ex-units 7376–1276–2251–1374)	Dean Forest Railway	York 1971
76811	DTCsoL	(ex-units 7390–1290–2255–1392)	Dean Forest Railway	York 1971
76817	DTCsoL	(ex-units 7396–1296–2258–1353)	Great Central Railway	York 1971
76818	DTCsoL	(ex-units 7397–1297–2256–1399)	Knights Rail Services, Eastleigh Works	York 1971
76833	DTCsoL	(ex-units 7412–1212–1881)	Barrow Hill Roundhouse	York 1971
76835	DTCsoL	(ex-units 7414–1214–1883–1497)	Mid Norfolk Railway	York 1971
76844	DTCsoL	(ex-units 7423–1223–1888–1498)	Swanage Railway	York 1972

62402 is named Freshwater.

▲ Restored to BR Green at the Great Central Railway, 4 Cig 7059 (76817+69339+62384+76246) is seen at Woodthorpe on 12 September 2009, being propelled by 33116 as the 10.00 Loughborough–Rothley. **Robert Pritchard**

▼ BR Blue 4 Vep 3417 (76262+70797+62236+76263) stands at the new Bluebell Railway East Grinstead station on 5 September 2010. **Phil Barnes**

CLASS 423 (4 Vep) BR

Built: 1967–74 for BR Southern Region.
System: 750 V DC third rail.
Formation: DTCsoL–MBSO–TSO–DTCsoL.
Traction Motors: Four English Electric EE507 of 185 kW.
Maximum Speed: 90 mph.

DTCsoL	20.18 x 2.82 m	35.0 tonnes	18/46
TSO	20.18 x 2.82 m	31.5 tonnes	–/98
MBSO	20.18 x 2.82 m	48.0 tonnes	–/76

62236	MBSO	(ex-units 7775–3075–3417)	Bluebell Railway	York 1969
70797	TSO	(ex-units 7717–3017–3417)	Bluebell Railway	Derby 1967
76262	DTCsoL	(ex-units 7717–3017–3417)	Bluebell Railway	York 1967
76263	DTCsoL	(ex-units 7717–3017–3417)	Bluebell Railway	York 1967
76875	DTCsoL	(ex-units 7861–3161–3545)	National Railway Museum, York (N)	York 1973
76887	DTCsoL	(ex-units 7867–3167–3568)	Woking Miniature Railway	York 1973

CLASS 489 (GLV) BR

Rebuilt: 1979–84 from Class 414 (2 Hap) as Motor Luggage Vans for the Victoria–Gatwick "Gatwick Express" service. Seats removed.
System: 750 V DC third rail.
Traction Motors: Two English Electric EE507 of 185 kW.
Maximum Speed: 90 mph.

DMLV	20.04 x 2.82 m	45 tonnes

61269–68500	GLV	(ex-units 6071–9101)	Ecclesbourne Valley Railway	Eastleigh 1959
61277–68503	GLV	(ex-units 6079–9104)	Spa Valley Railway	Eastleigh 1959
61280–68509	GLV	(ex-units 6082–9110)	Barry Rail Centre	Eastleigh 1959
61292–68506	GLV	(ex-units 6094–9107)	Ecclesbourne Valley Railway	Eastleigh 1959

CLASS 491 (4 TC) BR/METRO-CAMMELL

Built: 1967. Unpowered units designed to work push-pull with Class 430 (4 Rep) tractor units and Class 33/1, 73 and 74 locomotives. Converted from locomotive-hauled coaching stock built 1952–57 (original number in brackets).
Formation: DTSO–TFK–TBSK–DTSO.
Maximum Speed: 90 mph.

DTSO	20.18 x 2.82 m	32 tonnes	–/64
TFK	20.18 x 2.82 m	33.5 tonnes	42/–
TBSK	20.18 x 2.82 m	35.5 tonnes	–/32

70823 (34970)	TBSK	(ex-unit 412)	London Underground, West Ruislip Depot	MC 1957
70824 (34984)	TBSK	(ex-unit 413)	Midland Railway-Butterley	MC 1957
70826 (34980)	TBSK	(ex-unit 415)	Station Railway Heritage Centre, Sandford	MC 1957
70855 (13018)	TFK	(ex-unit 412)	Midland Railway-Butterley	Swindon 1952
70859 (13040)	TFK	(ex-unit 412)	Stravithie Station, Fife	Swindon 1952
70860 (13019)	TFK	(ex-unit 417)	Longstowe Station, near Bourn	Swindon 1952
71163 (13097)	TFK	(ex-unit 430)	London Underground, West Ruislip Depot	Swindon 1954
76275 (3929)	DTSO	(ex-unit 404)	St Leonards Railway Engineering	Eastleigh 1955
76277 (4005)	DTSO	(ex-unit 405)	Dartmoor Railway	Swindon 1957
76297 (3938)	DTSO	(ex-unit 415)	London Underground, West Ruislip Depot	Eastleigh 1955
76298 (4004)	DTSO	(ex-unit 415)	Midland Railway-Butterley	Eastleigh 1957
76301 (4375)	DTSO	(ex-unit 417)	The Heritage Centre, Bellingham	Swindon 1957
76302 (4382)	DTSO	(ex-unit 417)	The Heritage Centre, Bellingham	Swindon 1957
76322 (3936)	DTSO	(ex-unit 427)	Midland Railway-Butterley	Eastleigh 1955
76324 (4009)	DTSO	(ex-unit 428)	London Underground, West Ruislip Depot	Eastleigh 1957

76277 was also numbered DB977335 for a time when in departmental service.

CLASS 501 BR

Built: 1957 for Euston–Watford and North London lines.
Normal Formation: DMBSO–TSO–DTBSO.
System: 630 V DC third rail.
Traction Motors: Two English Electric GEC of 135 kW.
Maximum Speed: 60 mph.

DMBSO	18.47 x 2.82 m	47.8 tonnes	–/74
DTBSO	18.47 x 2.82 m	30.5 tonnes	–/74

BR	*AD*			
61183–DB 977349		DMBSO	Electric Railway Museum, Coventry	Eastleigh 1957
75186	WGP 8809	DTBSO	Electric Railway Museum, Coventry	Eastleigh 1957

CLASS 504 BR

Built: 1959 for Manchester–Bury line.
Normal Formation: DMBSO–DTSO.
System: 1200 V DC protected side-contact third rail.
Maximum Speed: 65 mph.
Traction Motors: Four English Electric EE of 90 kW.

DMBSO	20.31 x 2.82 m	50 tonnes	–/84
DTSO	20.31 x 2.82 m	33 tonnes	–/94

65451	DMBSO	East Lancashire Railway	Wolverton 1959
77172	DTSO	East Lancashire Railway	Wolverton 1959

TRAMS GRIMSBY & IMMINGHAM LIGHT RAILWAY

Built: 1925. (* 1927). Single-deck trams built for Gateshead & District Tramways Company. Sold to British Railways in 1951 for use on Grimsby & Immingham Light Railway.
Motors: 2 x Dick Kerr 31A (* 25A) of 18 kW (25 hp).
Bogies: Brill 39E.
Seats: 48 (* reduced to 44 by BR).

BR	*Gateshead*		
20 *	5	Crich Tramway Village	Gateshead & District Tramways Co. 1927
26	10	North of England Open Air Museum	Gateshead & District Tramways Co. 1925

BATTERY EMU BR DERBY/COWLAIRS TWIN UNIT

Built: 1958. Normal formation. BDMBSO–BDTCOL.
Power: 216 lead-acid cells of 1070 Ah.
Traction Motors: Two 100 kW Siemens nose-suspended motors.
Maximum Speed: 70 mph.

Vacuum brakes.

BDMBSO	18.49 x 2.79 m	37.5 tonnes	–/52
BDTCOL	18.49 x 2.79 m	32.5 tonnes	12/53

79998–DB 975003	BDMBS	Royal Deeside Railway	Derby/Cowlairs 1958
79999–DB 975004	BDTCL	Royal Deeside Railway	Derby/Cowlairs 1958

APPENDIX I. LIST OF LOCATIONS

The following is a list of locations in Great Britain where locomotives and multiple units included in this book can be found, together with Ordnance Survey grid references where these are known. At a small number of private sites the actual location of locomotives or rolling stock is not known – we would welcome any confirmation from readers. At certain locations locomotives may be dispersed at several sites, in such cases the principal site where locomotives can normally be found is the one given. Enquiries at this site will normally reveal the whereabouts of other locomotives or rolling stock, but this is not guaranteed.

PRESERVATION SITES & OPERATING RAILWAYS

§ denotes site not normally open to the public.

	OS GRID REF
Abbey View Disabled Day Centre, Neath Abbey, Neath Port Talbot.	SS 734973
Allely's Heavy Haulage, The Slough, Studley, Warwickshire.§	SP 057637
Aln Valley Railway, Longhoughton Goods Yard, Longhoughton, Northumberland	NU 240150
Appleby Heritage Centre, Appleby, Cumbria.	NY 688205
Appleby-Frodingham RPS, Tata Steel, Appleby-Frodingham Works, Scunthorpe, Lincs.	SE 913109
Avon Valley Railway, Bitton Station, Bitton, Gloucestershire.	ST 670705
Barclay Brown Yard, High Street, Methil, Fife.§	NT 372994
Barrow Hill Roundhouse, Campbell Drive, Staveley, Chesterfield, Derbyshire.	SK 414755
Barry Rail Centre, Barry Island, Vale of Glamorgan.	ST 118667
Battlefield Railway, Shackerstone Station, Shackerstone, Leicestershire.	SK 379066
Binbrook Trading Estate, Binbrook Airfield, near Louth, Lincolnshire.§	TF 201958
Birmingham Moor Street Station, Birmingham.	SP 074867
Black Bull Inn, Moulton, North Yorkshire.	NZ 237037
Bluebell Railway, Sheffield Park, near Uckfield, East Sussex.	TQ 403238
Boden Rail Engineering, Washwood Heath Depot, Birmingham.§	SP 103892
Bo'ness & Kinneil Railway, Bo'ness Station, Union Street, Bo'ness, West Lothian.	NT 003817
Bodmin & Wenford Railway, Bodmin General Station, Bodmin, Cornwall.	SX 074664
Bressingham Steam Museum, Bressingham Hall, near Diss, Norfolk.	TM 080806
Bryn Engineering, c/o Redrock Plant & Truck Services, Blackrod Industrial Estate, Scot Lane, Blackrod, near Bolton.§	SD 623089
Buckinghamshire Railway Centre, Quainton Road Station, Aylesbury, Bucks.	SP 736189
Caledonian Railway, Brechin Station, Brechin, near Montrose, Angus.	NO 603603
Cambrian Railway Trust, Llynclys, near Oswestry, Shropshire.	SJ 284239
Cefn Coed Colliery Museum, Old Blaenant Colliery, Cryant, Neath Port Talbot.	SN 786034
Chasewater Light Railway, Chasewater Pleasure Park, Brownhills, Staffordshire.	SK 034070
Chinnor & Princes Risborough Railway, Chinnor Cement Works, Chinnor, Oxon.	SP 756002
Cholsey & Wallingford Railway, St John's Road, Wallingford, Oxfordshire.	SU 600891
Churnet Valley Railway, Cheddleton Station, Cheddleton, Leek, Staffordshire.	SJ 983519
Cold Norton Play School, Palepit Farm, Latchingdon Road, Cold Norton, Essex.§	TL 858003
Colne Valley Railway, Castle Hedingham Station, Halstead, Essex.	TL 774362
Crewe Heritage Centre, Crewe, Cheshire.	SJ 708552
Crich Tramway Village, Crich, near Matlock, Derbyshire.	SK 345549
Darlington North Road Goods Shed, Station Road, Hopetown, Darlington, Co Durham.	NZ 290157
Dartmoor Railway, Meldon Quarry, near Okehampton, Devon.	SX 568927
Dartmouth Steam Railway, Queen's Park Station, Paignton, Devon.	SX 889606
Dean Forest Railway, Norchard, near Lydney, Gloucestershire.	SO 629044
Denbigh & Mold Junction Railway, Sodom, near Bodfari, Denbighshire, Wales.§	SJ 103711
Deptford Station Cafe, Deptford High Street, Greater London.	TQ 372773
Derwent Valley Light Railway, Yorkshire Museum of Farming, Murton, York, North Yorks.	SE 650524
Designer Outlet Village, Kemble Drive, Swindon, Wiltshire.	SU 142849
Didcot Railway Centre (Great Western Society), Didcot, Oxfordshire.	SU 524906
Ecclesbourne Valley Railway, Wirksworth Station, Wirksworth, Derbyshire.	SK 289542
East Anglian Railway Museum, Chappel & Wakes Colne Station, Essex.	TL 898289

East Kent Light Railway, Shepherdswell, Kent. TR 258483
East Lancashire Railway, Bolton Street Station, Bury, Greater Manchester. SD 803109
East Somerset Railway, West Cranmore Station, Shepton Mallet, Somerset. ST 664429
Eden Valley Railway, Warcop Station, Warcop, Cumbria. NY 753156
Electric Railway Museum, Coventry, Rowley Road, Baginton, Coventry, Warwickshire. SP 354751
Embsay & Bolton Abbey Railway, Embsay Station, Embsay, Skipton, North Yorks. SE 007533
Epping–Ongar Railway, Ongar Station, Station Road, Chipping Ongar, Essex. TL 552035
Fawley Hill Railway, Fawley Green, near Henley-on-Thames, Buckinghamshire.§ SU 755861
Ffestiniog Railway, Boston Lodge Works, Porthmadog, Gwynedd.§ SH 586380
Fighting Cocks PH, Middleton St George, near Darlington, Co Durham. NZ 342143
Finmere Station, near Newton Purcell, Oxfordshire.§ SP 629312
Flour Mill Workshop, Bream, Forest of Dean, Gloucestershire.§ SO 604067
Foxfield Railway, Blythe Bridge, Stoke-on-Trent, Staffordshire. SJ 976446
Garden Art, 1 Bath Road, Hungerford, Berkshire. SU 343689
Garw Valley Railway, Pontycymer Locomotive Works, Old Station Yard, Pontycymer, Bridgend. SS 904914
GCE & SCS, Tradecroft Industrial Estate, Easton, Isle of Portland, Dorset.§ SY 685723
Glasgow Riverside Museum, Pointhouse Quay, Yorkshill, Glasgow. NS 557661
Gloucestershire Warwickshire Railway, Toddington Station, Gloucestershire. SP 049321
Great Central Railway, Loughborough Central Station, Loughborough, Leicestershire. SK 543194
Great Yeldham, near Sudbury, Essex.§ TL 760379
Gwili Railway, Bronwydd Arms Station, Carmarthen, Carmarthenshire. SN 417236
Head of Steam, Darlington Railway Museum, North Road Station, Hopetown, Darlington, Co Durham. NZ 289157
Helston Railway, Trevarno Gardens, near Helston, Cornwall. SW 647306
Hever Station, Hever, Kent.§ TQ 465445
HM Prison Lindholme, Moor Dike Road, Lindholme, near Doncaster.§ SE 687057
Hope Farm (Southern Locomotives), Sellindge, near Ashford, Kent.§ TR 119388
Honeybourne Airfield Industrial Estate, Honeybourne, near Evesham, Worcs.§ SP 115422
Hydraulic House, West Bank, Sutton Bridge, near Spalding, Lincolnshire.§ TF 479209
Ian Storey Engineering, Station Yard, Hepscott, Morpeth, Northumberland.§ NZ 223844
Isle of Wight Steam Railway, Haven Street Station, Isle of Wight. SZ 556898
Keighley & Worth Valley Railway, Haworth, near Keighley, West Yorkshire. SE 034371
Keith & Dufftown Railway, Dufftown, Moray. NJ 323414
Kent & East Sussex Railway, Tenterden Town Station, Tenterden, Kent. TQ 882336
Kirklees Light Railway, Clayton West, near Huddersfield, West Yorkshire. SE 258112
Knights Rail Services, Eastleigh Works, Campbell Road, Eastleigh, Hampshire.§ SU 457185
Lakeside & Haverthwaite Railway, Haverthwaite, Cumbria. SD 349843
Lavender Line, Isfield Station, Station Road, Isfield, East Sussex. TQ 452171
Lincolnshire Wolds Railway, Ludborough Station, Ludborough, Lincolnshire. TF 309960
Little Mill Inn, Rowarth, Mellor, Derbyshire. SK 011890
Liverpool Museum Store, Juniper Street, Bootle, Merseyside.§ SJ 343935
Llanelli & Mynydd Mawr Railway, Cynheidre, near Llanelli, Carmarthenshire. SN 495071
Llangollen Railway, Llangollen Station, Llangollen, Denbighshire. SJ 211423
LNWR Crewe Carriage Shed, Crewe, Cheshire.§ SJ 715539
London Transport Depot Museum, Gunnersby Lane, Acton, Greater London. TQ 194799
London Underground, West Ruislip Depot, Ruislip, London.§ TQ 094862
Longstowe Station, near Bourn, Cambridgeshire.§ TL 315546
Mangapps Railway Museum, Southminster Road, Burnham-on-Crouch, Essex. TQ 944980
Marshalls Transport, Pershore Airfield, Throckmorton, near Evesham, Worcs.§ SO 978506
Mid Hants Railway, Ropley Station, Ropley, Hampshire. SU 629324
Middleton Railway, Tunstall Road, Hunslet, Leeds, West Yorkshire. SE 305310
Midland Railway-Butterley, Butterley Station, near Ripley, Derbyshire. SK 403520
Mid Norfolk Railway, Dereham Station, East Dereham, Norfolk. TF 994131
Moreton Park Railway, Moreton-on-Lugg, near Hereford, Herefordshire.§ SO 503467
Museum of Canterbury, Stour Street, Canterbury, Kent. TR 146577
Museum of Science & Industry, Liverpool Road, Castlefield, Greater Manchester. SJ 831978
National Railway Museum, Leeman Road, York, North Yorkshire. SE 594519
National Railway Museum, Shildon, Co Durham. NZ 238256
NELPG, Former Carriage Works, Hopetown, Darlington, Co Durham. NZ 288157
Nene Valley Railway, Wansford Station, Peterborough, Cambridgeshire. TL 093979
Northampton & Lamport Railway, Pitsford, Northamptonshire. SP 736666
Northamptonshire Ironstone Railway, Hunsbury Hill, Northampton, Northamptonshire. SP 735584
North Dorset Railway, Shillingstone Station, Shillingstone, Dorset. ST 824117

North Essex Traction Group, Hunnable Industrial Estate, Toppesfield Road, Great Yeldham, near Sudbury, Essex.§ TL 760379
North Norfolk Railway, Sheringham Station, Norfolk. TG 156430
North of England Open Air Museum, Beamish Hall, Beamish, Co Durham. NZ 217547
North Yorkshire Moors Railway, Grosmont Station, North Yorkshire. NZ 828049
Nottingham Transport Heritage Centre, Mereway, Ruddington, Nottinghamshire. SK 575322
Oswestry Railway Centre, Oswestry Station Yard, Oswestry, Shropshire. SJ 294297
Peak Rail, Rowsley South Station, near Matlock, Derbyshire. SK 262642
Plym Valley Railway, Marsh Mills, Plymouth, Devon. SX 520571
Pontypool & Blaenavon Railway, Furnoe Sidings, Big Pit, Blaenavon, Torfaen. SO 237093
Pullman Rail, Canton Depot, Cardiff.§ ST 177779
The Pump House Steam & Transport Museum, The Pump House, Lowe Hall Lane, Walthamstow, London. TQ 362882
Rail Restorations North East, Hackworth Industrial Park, Shildon, Co Durham.§ NZ 223255
Rampart Carriage & Wagon Services, London Road, Derby, Derbyshire.§ SK 362350
Reliance Industrial Estate, Eager Street, Newton Heath, Manchester.§ SD 873009
Ribble Steam Railway, off Chain Caul Road, Riversway, Preston, Lancashire. SD 504295
Rippingale Station, Fen Road, Rippingale, Lincolnshire.§ TF 115283
Rowden Mill Station, Rowden Mill, near Bromyard, Herefordshire.§ SO 627565
Royal Deeside Railway, Milton, Crathes, Banchory, Aberdeenshire. NO 743962
Rushden Transport Museum, Rectory Road, Rushden, Northamptonshire. SP 957672
Rutland Railway Museum, Ashwell Road, Cottesmore, Oakham, Rutland. SK 887137
RVEL, RTC Business Park, London Road, Derby, Derbyshire.§ SK 365350
Rye Farm, Ryefield Lane, Wishaw, Sutton Coldfield, Warwickshire.§ SP 180944
Science Museum, Imperial Institute Road, South Kensington, London. TQ 268793
Scolton Manor Museum, Scolton Manor, Haverfordwest, Pembrokeshire. SM 991222
Severn Valley Railway, Bridgnorth Station, Shropshire. SO 715926
Snibston Discovery Park, Snibston Mine, Ashby Road, Coalville, Leicestershire. SK 420144
South Devon Railway, Buckfastleigh, Devon. SX 747663
Southall Depot, Southall, Greater London.§ TQ 133798
Spa Valley Railway, Tunbridge Wells West Station, Tunbridge Wells, Kent. TQ 578385
St Leonards Railway Engineering, West Marina Depot, Bridge Way, St Leonards, East Sussex.§ TQ 778086
St Modwen Properties, Long Marston, Warwickshire.§ SP 152469
Stainmore Railway, Kirkby Stephen East Station, Kirkby Stephen, Cumbria. NY 769075
Station Railway Heritage Centre, Sandford, Somerset. ST 416595
Steam – Museum of the Great Western Railway, Old No. 20 Shop, Old Swindon Works, Kemble Drive, Swindon, Wiltshire. SU 143849
Stephenson Railway Museum, Middle Engine Lane, West Chirton, Tyne & Wear. NZ 323693
Stewarts Lane Depot, Battersea, Greater London.§ TQ 288766
Strathspey Railway, Aviemore, Highland Region. NH 898131
Stravithie Station, Stravithie, near St Andrews, Fife. NO 533134
Summerlee Museum of Scottish Industrial Life, West Canal Street, Coatbridge, North Lanarkshire. NS 728655
Swanage Railway, Swanage Station, Swanage, Dorset. SZ 028789
Swindon & Cricklade Railway, Blunsden Road Station, Swindon, Wiltshire. SU 110897
Talyllyn Railway, Pendre Depot, Tywyn, Gwynedd. SH 590008
Tanat Valley Light Railway, Nantmawr, near Llanyblodwel, Shropshire. SJ 253243
Tanfield Railway, Marley Hill Engine Shed, Sunniside, Tyne & Wear. NZ 207573
Tebay Depot, Cumbria.§ NY 614035
Telford Steam Railway, Bridge Road, Horsehay, Telford, Shropshire. SJ 675073
The Moor Business Park, Ellough Moor, near Beccles, Suffolk.§ TM 442885
The Heritage Centre, Bellingham, Northumberland. NY 843833
Thinktank: Birmingham Science Museum, Millennium Point, Curzon Street, Birmingham. SP 079873
Thornton Depot, Strathore Road, Thornton, Fife, Scotland.§ NT 264970
Titley Junction Station, near Kington, Herefordshire.§ SO 329581
Tiverton Museum, St Andrew's Street, Tiverton, Devon. SS 955124
Tyseley Locomotive Works, Warwick Road, Tyseley, Birmingham. SP 105841
Vale of Rheidol Railway, Aberystwyth, Ceredigian. SN 587812
Venice-Simplon Orient Express, Stewarts Lane Depot, Battersea, Greater London.§ TQ 288766
Waverley Route Heritage Association, Whitrope, near Hawick, Scottish Borders. NT 527005
Weardale Railway, Wolsingham, Co Durham. NZ 081370
Welshpool & Llanfair Light Railway, Llanfair Caereinon, Powys. SJ 107069

Wembley Depot (Alstom), Wembley, Greater London.§	TQ 193844
Wensleydale Railway, Leeming Bar Station, Leeming Bar, North Yorkshire.	SE 286900
West Coast Railway Company, Warton Road, Carnforth, Lancashire.§	SD 496708
West Somerset Railway, Minehead Station, Minehead, Somerset.	SS 975463
Whitwell & Reepham Station, near Alysham, Norfolk.§	TG 091217
Willesden Depot (London Overground), Willesden, Greater London.§	SE 200840
Woking Miniature Railway, Barr's Lane, Knaphill, Woking, Surrey.	TQ 966595
Yeovil Railway Centre, Yeovil Junction, near Yeovil, Somerset.	ST 571141

APPENDIX II. ABBREVIATIONS USED

ABB	ASEA Brown Boveri
AC	Alternating Current
AD	Army Department of the Ministry of Defence
AD	Alexandra Docks & Railway Company
BAD	Base Armament Depot
BP	BP (formerly British Petroleum)
BR	British Railways
BPGVR	Burry Port & Gwendraeth Valley Railway
BREL	British Rail Engineering Ltd. (later BREL, then ABB, now Bombardier Transportation).
BTH	British Thomson Houston
BUT	British United Traction
CARR	Cardiff Railway
CR	Caledonian Railway
DC	Direct Current
EE	English Electric Company
FR	Furness Railway
FS	Ferrovie dello Stato (Italian State Railways)
GCR	Great Central Railway
GEC	General Electric Company (UK)
GER	Great Eastern Railway
GJR	Grand Junction Railway
GNR	Great Northern Railway
GNSR	Great North of Scotland Railway
GSWR	Glasgow & South Western Railway
GVR	Gwendraeth Valley Railway
GWR	Great Western Railway
H&B	Hull & Barnsley Railway
HR	Highland Railway
L&MR	Liverpool & Manchester Railway
LBSCR	London, Brighton & South Coast Railway
LCDR	London, Chatham & Dover Railway
LMS	London Midland & Scottish Railway
LNER	London & North Eastern Railway
LNWR	London & North Western Railway
LSWR	London & South Western Railway
L&Y	Lancashire & Yorkshire Railway
LTE	London Transport Executive
LTSR	London Tilbury & Southend Railway
MAV	Hungarian Railways
MR	Midland Railway
MoD	Ministry of Defence
MSJ&A	Manchester South Junction & Altrincham Railway
NBR	North British Railway
NELPG	North Eastern Locomotive Preservation Group
NER	North Eastern Railway
NLR	North London Railway
NS	Nederlandse Spoorwegen (Netherlands Railways)
NSR	North Staffordshire Railway
P&M	Powlesland & Mason

PKP	Polish Railways
PTR	Port Talbot Railway
RLC	Royal Logistics Corps
ROD	Railway Operating Department
RPS	Railway Preservation Society
S&DJR	Somerset & Dorset Joint Railway
SDR	South Devon Railway
SECR	South Eastern & Chatham Railway
SER	South Eastern Railway
SJ	Swedish State Railways
SPR	Sandy & Potton Railway
SR	Southern Railway
TCDD	Türkiye Cumhuryeti devlet Demiryollan (Turkish Railways)
TVR	Taff Vale Railway
USATC	United States Army Transportation Corps
W&L	Welshpool & Llanfair Railway
WD	War Department
WR	British Railways Western Region
WT	Wantage Tramway

APPENDIX III. PRIVATE MANUFACTURER CODES

The following codes are used to denote private locomotive manufacturers. These are followed by the works number and build year, eg AW 1360/1937 – built by Armstrong-Whitworth and Company, works number 1360, year 1937. Unless otherwise shown, locations are in England.

AB	Andrew Barclay, Sons & Company, Caledonia Works, Kilmarnock, Scotland.
AC	AC Cars, Thames Ditton, Surrey.
AE	Avonside Engine Company, Bristol, Avon.
AEC	Associated Equipment Company, Southall, Berkshire.
AL	American Locomotive Company, USA/Canada.
AW	Armstrong-Whitworth & Company, Newcastle, Tyne & Wear.
BBC	Brown-Boveri et Cie., Switzerland.
BCK	Bury, Curtis & Kennedy, Liverpool, Merseyside.
BE	Brush Electrical Engineering Company, Loughborough, Leicestershire.
BLW	Baldwin Locomotive Works, Philidelphia, Pennsylvania, USA.
BMR	Brecon Mountain Railway Company, Pant, Merthyr Tydfil, Wales.
Boston Lodge	Ffestiniog Railway, Boston Lodge Works, Porthmadog, Wales.
BP	Beyer Peacock and Company, Gorton, Manchester.
BRCW	Birmingham Railway Carriage & Wagon Company, Smethwick, Birmingham.
BTH	British Thomson-Houston Company, Rugby, Warwickshire.
CE	Clayton Equipment Company, Hatton, Derbyshire.
Cravens	Cravens, Darnall, Sheffield, South Yorkshire.
Darlington Hope Street	A1 Steam Trust, Darlington Hope Street, Darlington, Co Durham.
DC	Drewry Car Company, London.
Didcot Railway Centre	Great Western Society, Didcot, Oxfordshire.
DK	Dick Kerr & Company, Preston, Lancashire.
Dodman	Alfred Dodman & Company, Highgate Works, Kings Lynn, Norfolk.
EE	English Electric Company, Bradford and Preston.
FW	Fox, Walker & Company, Atlas Engine Works, Bristol.
Gateshead & District Tramways Co,	Sunderland Road Works, Gateshead, Tyne & Wear.
GCR Dukinfield	Great Central Railway Carriage & Wagon Works, Dukinfield, Tameside, Greater Manchester.

GE	George England & Company, Hatcham Ironworks, London.
GRCW	Gloucester Railway Carriage & Wagon Company, Gloucester, Gloucestershire.
Hack	Timothy Hackworth, Soho Works, Shildon, Co. Durham.
HC	Hudswell-Clarke & Company, Hunslet, Leeds, West Yorkshire.
HE	Hunslet Engine Company, Hunslet, Leeds, West Yorkshire.
HL	R&W Hawthorn, Leslie & Company, Forth Bank Works, Newcastle-upon-Tyne.
HLT	Hughes Locomotive & Tramway Engine Works, Loughborough, Leicestershire.
K	Kitson & Company, Airedale Foundry, Hunslet, Leeds, West Yorkshire.
Kitching	A Kitching, Hope Town Foundry, Darlington, Co Durham.
KS	Kerr Stuart & Company, California Works, Stoke-on-Trent, Staffordshire.
Leyland	British Leyland Lillyhall Works, Workington, Cumbria.
Lima	Lima Locomotive Works Inc., Lima, Ohio, USA.
Llangollen Railway	Llangollen Railway, Llangollen, Clwyd.
Loco. Ent.	Locomotion Enterprises (1975), Bowes Railway, Springwell, Gateshead, Tyne & Wear.
Manch	Museum of Science & Industry, Liverrpool Road, Manchester.
MC	Metropolitan-Cammell Carriage & Wagon Company, Birmingham (Metro-Cammell).
MV	Metropolitan-Vickers, Trafford Park, Manchester.
N	Neilson & Son, Springburn Locomotive Works, Glasgow, Scotland.
NBL	North British Locomotive Company, Glasgow, Scotland.
NR	Neilson Reid & Company, Springburn Works, Glasgow, Scotland.
PR	Park Royal Vehicles, Park Royal, London.
PS	Pressed Steel, Swindon, Wiltshire.
Resco	Resco (Railways), Erith, London.
RH	Ruston & Hornsby, Lincoln.
RS	Robert Stephenson & Company, Newcastle, Tyne & Wear.
RSH	Robert Stephenson & Hawthorns, Darlington, Co Durham.
RTC	Railway Technical Centre, Derby, Derbyshire.
S	Sentinel (Shrewsbury), Battlefield, Shrewsbury, Shropshire.
Sara	Sara & Company, Plymouth, Devon.
Science Museum	Science Museum, South Kensington, London
SM	Siemens, London.
SS	Sharp Stewart & Sons, Manchester and then (1888) Glasgow.
TKL	Todd, Kitson and Laird, Leeds, West Yorkshire.
VF	Vulcan Foundry, Newton-le-Willows, Lancashire.
VIW	Vulcan Iron Works, Wilkes-Barre, Philadelphia, Pennsylvania, USA.
WB	WG Bagnall, Castle Engine Works, Stafford, Staffordshire.
Wkm	D Wickham & Company, Ware, Hertfordshire.
WMD	Waggon und Maschienenbau GmbH, Donauworth, Germany.
WSR	West Somerset Railway, Minehead, Somerset.
YE	Yorkshire Engine Company, Meadowhall, Sheffield, South Yorkshire.